D

EVERYTHING CATS
EXPECT YOU
TO KNOW

EVERYTHING CATS
EXPECT YOU
TO KNOW

Elizabeth Martyn

Good Books

Intercourse, PA 17534
800/762-7171
www.GoodBooks.com

EVERYTHING CATS EXPECT YOU TO KNOW

International Standard Book Number: 978-1-51648-625-0
Library of Congress Catalog Card Number: 2008012983

Text copyright © 2007 by Elizabeth Martyn
Artworks reproduced with permission from Dover Publications, Inc.
Original edition copyright © 2007 by New Holland Publishers (UK) Ltd.

Printed in the United States of America

Library of Congress Cataloging-in-Publication Data

Martyn, Elizabeth.
  Everything cats expect you to know / Elizabeth Martyn.
     p. cm.
  Includes index.
  ISBN 978-1-56148-625-0 (hardcover : alk. paper) 1. Cats--Miscellanea. I. Title.
  SF445.5.M387 2008
  636.8--dc22   2008012983

CONTENTS

INTRODUCTION

Cats are one of life's great pleasures, with their caressable

coats, unselfconscious beauty and delightful little ways.

Bringers of laughter and lightness to everyday life, cats are

necessary to my well-being – and maybe yours, as well –

and writing about them in this book has been thoroughly

enjoyable.

I've been lucky enough to share my life with some charming cats over the years. There's always been a tabby around the house, from fluffy Tinky to Maisie, a cute little rescue puss; Bibi, who would do anything for a bowl of raw liver; Sophie, grande dame of tabbies, who thrived for 20 years thanks to her motto, "Take no risks"; and Rosie, a three-legged, pocket-sized mackerel tabby with a huge personality. Along the way, there's also been Ambrose, a lithe ginger tom, and the current cat-in-residence, Poppy, a sweet-natured tortoiseshell with white bib and paws.

And Poppy it is who's sat beside me while I've been writing this book. She's inspected cats on the computer screen, lounged contentedly amid my mountain of books, looked puzzled when I've chuckled over feline antics on YouTube, and crept comfortingly on to my lap when she caught me sniffing over tales of the Rainbow Bridge. When I've been very engrossed, she's even sneaked into the forbidden zone under the desk lamp until, alerted by the unmistakable aroma of scorched fur, I've sent her off to shoehorn herself between the kids to watch TV, or insinuate herself into that enticing cranny between husband and newspaper.

I hope you enjoy reading this collection of everything your cat could possibly want you to know – and a lot more besides – as much as I've enjoyed compiling it. You can see me and Poppy at work, and read more about the creation of this book, at my Web site, www.wordswoman.com/cats.html

Elizabeth Martyn

✒ ARE BLACK CATS LUCKY ... OR UNLUCKY? ✒

That depends on where you live – and when you lived.

Between the superstitious medieval ages and the 17th century, black cats were feared across Europe. The color black was strongly associated with Satan, and people thought that black cats were impish familiars, and consorted with demons. Old women who cared for black cats were in danger of being branded witches, and thrown to the flames, along with their cats.

Over time, beliefs changed. Black cats were still seen as powerful, but now their skill lay in keeping the Devil at arm's length, and they turned into symbols of good luck.

To be truly lucky, you should meet a black cat by accident – no lying in wait for your luck to wander past. It's even better if one crosses your path – although only from right to left – and you can strengthen the luck by stroking the cat three times. Once you've done that, leave quickly, because all the luck is lost if the cat turns back on itself or bolts.

Black cats are still bad news in the United States, Italy and Portugal, where the rarer white breeds are the luck-bringers. If you see a black cat when you're out for a walk in Bulgaria, hop around in a circle, in three hops. Otherwise woe will betide you.

And stay away from Sooty in China, where black cats portend illness and poverty.

ALBERT SCHWEITZER (1875–1965), MISSIONARY SURGEON

Schweitzer had many talents. He was not only a philosopher, physician and theologian, but a musician, author and builder as well. He was awarded the Nobel Peace Prize for his work at the Schweitzer Hospital at Lambarene, Gabon, in Equatorial Africa, where he lived surrounded by his beloved animals, whose comings and goings he noted in his diary.

Schweitzer loved and valued all animals. His philosophy was respect for all life, and he would even scoop toads and insects out of holes dug during building work, so that no living creature would be injured or killed.

Along with numerous dogs, pigs, monkeys, chimps, gorillas, pelicans and an antelope, the Lambarene complex was inhabited by cats.

Sizi was trapped under the floor of a newly-constructed building as a kitten, and rescued by Schweitzer, who heard her panic-stricken meow. She lived to be 23, and liked to sit beside her savior when he was writing, and would sometimes drop off with her head on his left arm, which the gentle doctor wouldn't move until she woke.

Another cat, Piccolo, liked to indulge in a siesta on top of the papers piled on Dr. Schweitzer's desk for signature.

What do you get if you cross
a cat with a gorilla?
A pet that puts you out at night!

DO CATS LIKE SINKS?

Oh, yes, they do. It might be the size of a wash basin, which a cat fits into so snugly. Maybe it's the shape, rounded and comfy. Perhaps it's the shine, always an appealing factor, or the fun of tinkering with sparkly taps or fighting off the plug chain.

Whatever it is, once your cat has discovered the joys of bathroom life, you'll go to brush your teeth and find the sink is already occupied.

There's even a Web site devoted to the subject, at http://catsinsinks.com. With several hundred images, and thousands more to come, devoted owners worldwide have uploaded pictures of, um, their cats. In sinks.

❧ CATS THAT APPEAR IN THE JAMES HERRIOT BOOKS ❧

Many cats were portrayed in these popular "vet" books.
Read more about the series on page 14.

ALFRED

Lived in the candy store. He lost weight, and had to
have a hairball surgically removed.

BORIS

A ferocious outdoor cat and cunning escape artist.

EMILY

A small cat, who gave birth to a very large kitten.

FRISK

Had everyone worried when he kept collapsing. It
turned out he was drinking his master's medicine.

GINNY AND OLLIE

A pair of feral cats who hung around the vet's house, but
refused to come inside. They preferred Herriot's wife
because the vet kept trying to give them nasty treatments.

GEORGE

A feral cat, whose inner eyelid was paralyzed.

MOSES

So named because he was found in the rushes, Moses
was a black tom kitten who accepted a sow, Bertha, as his
foster mother and suckled from her along with her piglets.

OSCAR AKA TIGER

Large, sociable and exceedingly stripy, this tabby took
turns living with different families.

MAMMA MIA – WHOSE CAT ARE YOU?

Freddie Mercury (1946–1991), lead singer and pianist with Queen, composed "Bohemian Rhapsody" and was a flamboyant performer who thrived on being on stage.

Offstage, he adored cats, and owned as many as 10 at a time, including Tom, Jerry, Oscar, Tiffany, Delilah, Goliath, Miko, Romeo and Lily. He sometimes wore clothes picturing cats when he was photographed for album covers and in videos.

Mercury's personal assistant, Peter Freestone, thought his boss "put as much importance on them [his cats] as any human life." Mercury thought of his cats as his children, and when Mary Austin, his girlfriend – he was bisexual – suggested they have a baby, he said: "I'd rather have another cat."

He would call home from concert tours and baby-talk to his cats, who comforted him by lying close to him on his bed when he was dying of AIDS.

Ann Ortman, a painter, was invited to meet Mercury and the five felines he then owned so that she could paint the cats' portraits. There were framed photos of Mercury's cats on every surface in the room where they met. Ann spent an afternoon photographing the cats in different positions, and "inevitably, their anxious owner, almost like a proud parent, crept into the pictures and I'm pleased to say that I have some wonderful photographs of that day to remind me of Freddie, especially ones of him sitting with his cats on his lap."

He dedicated both his album, *Mr. Bad Guy* and the song "Delilah" to cats.

How do you know if your cat's caught a duckling?
She looks down in the mouth.

✦ SHOULD I ASK MY CAT TO MY WEDDING? ✦

If you can persuade a black cat to cross the newlyweds' path as they leave the church, they are guaranteed a happy life together.

In France, a wet wedding day was the reward of a bride who was cruel to cats, while finding a white cat at the door on the wedding morning meant the marriage would be a success. Not so in the Netherlands, where cats had to be kept away from the door before the wedding to avoid bad luck and a disrespectful wife. If a cat did approach the bride, all would be well – as long as it sneezed.

✦ HITTING THE HIGH NOTES ✦

French author and cat fanatic Theodore Gautier (1811–1872) cared for cats above everything, and owned many, including Don Pierrot de Navarre, who would perch on his desk and follow the movement of his hand as he wrote.

His ginger-and-white cat, Madame Théophile, was a knowledgeable music critic. "Sitting on a pile of scores she listened attentively and with visible signs of pleasure to singers. But piercing notes made her nervous and at the high A she never failed to close the mouth of the singer with her soft paw. This was an experiment which it amused many to make and which never failed. It was impossible to deceive this cat dilettante on the note in question."

> *"The cat, having sat upon a hot stove lid,*
> *will not sit upon a hot stove lid again.*
> *Nor upon a cold stove lid."*
>
> Mark Twain, American author of
> *The Adventures of Huckleberry Finn*

✦ WHO SENT A FELINE PHALANX INTO BATTLE? ✦

THE BATTLE OF PELUSIUM, 525 B.C.

Cambyses of Persia had a savage, ugly temper, but he was a clever and skillful soldier. When he embarked on his attack on Africa, he first led his army safely across the deserts that lay between Persia and Egypt, and arrived outside the city walls of Pelusium in 525 B.C.

There, Cambyses unveiled his simple but very imaginative secret weapon. Knowing the cat was sacred in Egypt, he ordered his men to strap live cats to their shields, then advance. The Egyptians had no choice but to lay down their bows and arrows because, for them, to injure a sacred animal was a serious offense. Legend says that the battle was won without a blow being struck, but this can't be strictly true because later the Greek author Herodotus (c. 480–429) examined the bones of the two opposing sides (and noted that the Egyptians had thick skulls, while the Persian skulls were thin, a fact he put down to the Egyptian habit of head-shaving).

Whatever the exact facts, the Persians certainly stormed the city. After the battle, Cambyses decided to show his contempt for the Egyptians' cowardice, and rode around the city on a horse with a cage of cats, which he then flung at his enemies with taunts and insults.

This was just one of the many insults Cambyses directed against Egyptian religion. He eventually went mad.

❧ TIME SPENT WITH CATS IS NEVER WASTED ❧

The French author Colette adored cats, often wrote about them and owned many during her lifetime, including Kiki-la-Doucette, a gray Angora, featured as a talking cat in her book *Sept Dialogues des Bêtes* (1903). She also kept an African wild cat, Batou, for a time, but reluctantly sent him to the zoo, when she noticed his unhealthy interest in her puppy.

❧ THINGS KIDS SAY ABOUT CATS ❧

"Never hold a cat and a dust-buster at the same time."
Kyoya, 9

"If you want a pet kitten, start off by asking for a pony."
Naomi, 15

❧ IT SHOULDN'T HAPPEN TO A VET ❧

"I have felt cats rubbing their faces against mine and touching my cheek with claws carefully sheathed. These things, to me, are expressions of love."

These are the words of Alf Wight, born in 1916, and better known as best-selling veterinarian-cum-author James Herriot. Herriot didn't start writing until he was 50. "For years I used to bore my wife over lunch with stories about funny incidents. The words 'My book,' as in 'I'll put that in it one day,' became a sort of running joke. Eventually she said, 'Look, I don't want to offend you, but you've been saying that for 25 years.… Old vets of 50 don't write books.' So I purchased a lot of paper right then and started to write."

The result was a series of books which were enormously successful in the United Kingdom and the United States, plus two films and a long-running BBC television series, *All Creatures Great and Small.*

The books, which Herriot referred to as his "little cat-and-dog stories," document a time when veterinary practice was moving from the ancient remedies still used to antibiotics. Herriot's sharply observant ear and eye picked up on the suspicions of his farming clientele, some of whom were highly skeptical about his newfangled treatments, and preferred to stick with their own superstitious ways of treating sick animals.

Animal medicine was still fairly primitive, and among the stories in the book are Wight's recollections of performing his first hysterectomy on a cat and his first (nearly disastrous) abdominal surgery on a cow.

⊛ RECORD-BREAKING *CATS* ⊛

When Andrew Lloyd Webber adapted T. S. Eliot's whimsical poems, *Old Possum's Book of Practical Cats* into the musical *Cats*, he could never have guessed that a show about the antics of a bunch of street mogs would create creating a record-breaking, award-winning show. With numbers like "Jellicle Songs for Jellicle Cats," "Grizabella: The Glamour Cat," "Mr. Mistoffelees" and, by far the best-known, "Memory," the musical captured the public's imagination. It has had record-breaking runs in London and New York, been translated into 20 languages, won awards galore and been staged countless times in theaters all over the world.

Cats opened in London on May 11, 1981, and ran for nearly 9,000 performances before closing exactly 21 years later, on May 11, 2002. It ran on Broadway for more than 7,000 performances, and only lost its title of longest-running musical in 2006, when the record was taken by *Phantom of the Opera*, another Andrew Lloyd Webber musical.

⊛ HOW A CAT INSPIRED A HORROR STORY ⊛

Edgar Allen Poe, 19th-century American author, used the loyalty of his tortoiseshell cat, Catarina, as inspiration for his horror story, "The Black Cat," in which a ghostly cat betrays a murderer. Poe couldn't afford proper heating for his wife, Virginia, when she lay dying of tuberculosis during the winter of 1846. He wrapped her in his great coat and placed Catarina beside her. The cat snuggled up to the sick woman for warmth and "seemed conscious of her great usefulness."

CAT NAMES INSPIRED BY DRINKS

Brandy	Coke	Pernod
Burgundy	Cola	Sauvignon
Buttermilk	Daiquiri	Scotch
Cappuccino	Earl Grey	Soda
Chablis	Guinness	Tequila
Champagne	Milkshake	Vodka
Chardonnay	Mr. Pepsi	Whisky
Coffee	Pepsi	

WHO WROTE THE FIRST CAT POEM IN EUROPEAN LITERATURE?

That well known poet "Anon," in this case a ninth-century Irish monk, doodling idly in the margins of a manuscript and watching the antics of his pet cat, Pangur Bán (White Cat), while he probably should have been working.

PANGUR BAN

I and Pangur Ban, my cat,
'Tis a like task we are at;
Hunting mice is his delight,
Hunting words I sit all night.

Better far than praise of men
'Tis to sit with book and pen;
Pangur bears me no ill will;
He, too, plies his simple skill.

'Tis a merry thing to see
At our task how glad are we,
When at home we sit and find
Entertainment to our mind.

Oftentimes a mouse will stray
Into the hero Pangur's way;
Oftentimes my keen thought set
Takes a meaning in its net.

'Gainst the wall he sets his eye
Full and fierce and sharp and sly;
'Gainst the wall of knowledge I
All my little wisdom try.

When a mouse darts from its den.
O how glad is Pangur then!
O what gladness do I prove
When I solve the doubts I love!

So in peace our tasks we ply,
Pangur Ban, my cat and I;
In our arts we find our bliss,
I have mine, and he has his.

Practice every day has made
Pangur perfect in his trade;
I get wisdom day and night,
Turning Darkness into light.

❧ DO CATS HAVE TO HUNT? ❧

You can't deny what's in the genes, and the hunting instinct will show itself even in the cuddliest, best-fed feline.

But although it's natural, hunting is a destructive trait. The average domestic cat puts an end to 40 small creatures every year, and some rack up far more than that. Too many of the victims are garden birds, whose numbers are already endangered. Although the effect on bird populations of cat hunting is peanuts in comparison to the threat posed by destruction of their habitat, you should still take steps to minimize your cat's impact on the local wildlife.

So deep is the drive to stalk and pounce that cats will do it even if they're not hungry. One biologist gave cats a bowl of their favorite food, then let a rat loose in the room. Result? Even a helping of lightly-poached fish isn't enough to keep a cat from killing. The cats would stop eating for long enough to dispatch the rat, but rather than consume the prey would go back to their abandoned dinner. So, although feeding your cat well may discourage her from eating what she catches, it probably won't entirely dissuade her from hunting.

When an urban cat is ravenous, it's more likely to scavenge for food than hunt for it. It's a lot easier to rip open a bag than to patiently stalk a tiny mouse, and the pickings are richer, too. Feral cats soon learn which back doors on a street belong to restaurants and supermarkets.

❧ HOW MANY CATS SHOULD I HAVE? ❧

People love the idea of two – or more – cats. Cats are less enthusiastic, and many prefer to keep the house – and their owners – strictly to themselves.

In practical terms, whether you take on one kitten or a couple doesn't make much difference. You'll spend more on food, vet's bills and boarding (if you need it). But if your budget allows, then that's not a big problem. Your furniture will be that much furrier, but on the plus side you'll have the twice the amount of pleasure from stroking and playing with your pets, and watching them together.

If you bring home two kittens from the same litter, who have never lived alone, then you shouldn't have a problem, and they may even grow up to be best buddies and curl up together in a basket every night. Be prepared, though, for one or other of them to behave as if their housemate is an unpleasant smell. Even siblings can grow up having regular spats and barely tolerating each other.

Let's say you already have an adult cat who's well established in your home. Bubbles is such a delight that you can't think of anything nicer than having another cat. Double the pleasure, right?

But Bubbles may be so put out by the arrival of Squeak that she tries to leave home, becomes aggressive or tries a few little tricks like pooping on the carpet just to let you know, "I want to be alone."

It's usually easier to introduce a kitten than an adult into a household where there's already a cat. Bubbles should find a kitten less threatening, and a kitty will respond to her playfully rather than aggressively, and may even win her over.

It might be slightly easier to introduce a male cat, if you already own a neutered female. In the wild, female cats chase off other females fiercely, but are happy for males to come and go as they please. If your present incumbent is a neutered male, in theory he should – eventually – accept a cat of either sex. That's in theory.

Collecting cats can become quite a habit, and there are plenty of instances of people keeping large numbers together. Cats don't show stress readily though, and will often keep their angst internalized by huddling themselves away, hiding or staying out all the time. Keep a close eye on your cats, and don't strain them by introducing more and more pets if things are working well as they are.

❧ YOU KNOW YOU'RE A CAT PERSON WHEN ... ❧

… You call the bathroom the "tray."

… You don't feel dressed without a fine layer of cat hair.

… You treat fur in your food as extra fiber.

… You say "Sorry" when you step on a cat toy in the dark.

… You pat the sofa beside you when you invite a guest to sit down.

… You sleep clinging to the edge of the bed because your cat looks soooo sweet spread-eagled across the middle.

… You absent-mindedly put your child's dinner plate on the floor.

… You spent more on cat toys than kids' toys at Christmas.

… Your neighbors refer to you as "the crazy one with all the cats."

… You carry more photos of your cats, than of your kids.

… You keep calling your partner "Smokey."

… You'll only be friends with people if your cat likes them.

… You watch bad TV shows because Smokey is sleeping on the remote.

… Your cat sleeps on your head. And you like it.

… When people phone you, you insist they have a little chat with your cat as well.

… When there's a new visitor, you introduce them to your cat by name.

… You set a place for your cat at the table – or ON the table.

… Your answering machine message ends with "meow."

… Your partner says, "It's me or the cat," and you don't hesitate.

❦ TELL ME A CAT SPELL … ❦

- No part of the cat's body was spared in medieval times, when concoctions prepared from the dried liver, blood and bones were used to treat afflictions such as blindness, skin diseases, and even unrequited love and depression.
- To render yourself invisible, hold the bone of a black cat beneath your tongue.
- Cure styes by stroking the tail of a live black cat back and forth across the affected eye.
- Spots, sores or nettle rash? Rub gently with a cat's tail. The tail can also be used to treat epilepsy.
- A witch can turn herself into a cat, by rubbing on an ointment made from the fat of a fine black cat.

❦ MENDELSSOHN'S DISAPPOINTMENT ❦

Writing in the *Saturday Review* in 1904, John F. Runciman declares that his cats, among them one named Felix Mendelssohn, loved a family sing-along. "I have known them even to like such songs as … 'The Horse What Missus Dries the Clothes On,' and 'The Boers Got My Daddy' … but it is necessary to close the top of the piano or you may find the instrument clogged with bits of meat, dead mice, corks, etc."

The cats, he believed, were disappointed that they couldn't extract a few notes out of a violin bow, in the same way as they could with the piano keyboard. "Though they can knock a bow on the floor and shove it about, nothing in the way of music comes of it. Mr. Balling once played the viola-alta at my house and the eyes of Felix Mendelssohn glistened with hope. The performance over the bow was duly experimented with. Alas! – no result. And Felix retired to a corner and sat there half an hour wrapt in melancholy thought."

❧ THE THINGS CATS CATCH ❧

Most cat owners have been presented with their pets' prey. It's part of cat-owning, and you soon get used to disposing of the remains of half-eaten mice. In fact, it's always worse if the prey comes back to life after the cat has lost interest. Having a half-dead gull flapping around the kitchen is no joke, and just try coaxing a squirrel out from under the Aga.

Here are some owners' tales of memorable feline trophies.

Jane: "Cosmo came home making the triumphant hunter's mouth-filled yowl. Out of the corner of my eye, I saw her carrying a large, dark, rat-shaped victim. Shock, horror, terror, trepidation. It turned out to be an avocado."

Robbie: "Megan regularly brought home slices of bread that a neighbor had put out for the birds."

Tim: "There's a butcher's shop 12 doors down – I counted. Bibi struggled in through
the cat flap one night dragging a muddy pig's trotter (foot). I haven't a clue how she got it over 12 fences."

Kate: "Willum trotted in one night with one of those bread rolls shaped like a mini-Hovis loaf stuffed in his mouth. He could hardly hold it. His first-ever catch was a cauliflower stump, which he kept in his basket and played with for several weeks."

Leah: "I came home one evening and as soon as I opened the door I could hear crashing, and see feathers. It was a terrified blackbird, hurling itself at the … window. Sparky was in her basket, trying to look nonchalant."

❧ WHAT'S A CAT'S MOTTO? ❧

No matter what you've done – make it look like the dog did it.

⚉ CAN CATS FLY? ⚉

- Tiger the tabby lived on a German army base, and decided to take a nap in a crate one afternoon. As he slumbered peacefully, the crate was stowed on a plane, and Tiger woke up in Northern Ireland. When the story made the local news, people contributed to his quarantine fees, and soon he was flown back home.

- Cudzoo almost took off when she fell out of the 20th-floor window of a Manhattan skyscraper. It was pure luck than an awning was open further down the building, which provided a soft and safe landing.

- In 1919, the super-zeppelin *R34* was the first ever aircraft to cross the Atlantic from east to west. On board were eight officers, 22 crew, one human stowaway – and Wopsie, a cat smuggled aboard to bring the venture good luck. The trip took 108 hours, and the ship survived icy temperatures, strong headwinds and a thunderstorm before landing on Long Island.

⚉ WHAT DO CATS WATCH ON TV? ⚉

An Arizona-based online radio and TV station, Cat Galaxy claims to be designed especially "for felines to listen to and watch," and airs films such as *In the Cat's Eyes*, which features a large cat staring solemnly out of a window – an activity which most cats find more absorbing than watching television.

The station keeps its feline audiences and their owners happy 24/7, with programs such as *Morning Meows, Meow Mixing Monday, Tuesday Night Cat Club, Wednesday Night Cat Attack, Thursday Night Purr Party* and the *Friday Night Feline Frenzy*. Cat Galaxy also campaigns for cat charities, and airs details from animal shelters of cats that need homes.

Why did the cat train as a paramedic?
She wanted to be a first-aid kit!

❧ HOW DO YOU CHOOSE A KITTEN? ❧

If you prefer a pedigree, find out about the breed's characteristics first. Some purebred cats tend to be neurotic and noisy. Go to a reputable breeder for your kitten.

If you're happy with a non-pedigree cat, try local cat rescue programs. Be aware that potential owners are carefully checked by reputable rescuers, so be prepared for an interview and home visit before you're allowed to choose a pet.

Refuse to be seduced by an adorably fuzzy little ball of fluff unless you're sure you can commit to regular grooming. Longhaired cats will take hours of your elbow grease to keep their coats tangle-free, while shorthairs are virtually maintenance-free.

When you go to select your kitty:

- Look for a glossy coat, clear skin, bright eyes, clean nose and ears, no signs of an upset stomach, strong white teeth.
- Choose a kitten that's playful and energetic, not too timid, able to leap and run with no problems.

Don't take a kitten from its mother until it's at least 10 weeks old. Make sure she has had the necessary vaccinations before you let her outside, and consider having her fitted with a security chip, especially if she's a valuable – and stealable – pedigree. Before kitty moves in, you'll need:

- a cat bed
- litter tray, scoop and supply of cat litter
- food and water bowls
- brush and comb if she's longhaired
- a few simple playthings

If your new pet is going to be allowed out, consider fitting a cat flap. Lockable ones give you some control, and mean she can be kept in – and marauders can be kept out – overnight. Or you can buy a cat flap which is opened by a magnet attached to the cat's collar, although some cats are frightened of these and find it hard to learn how to use them or don't like wearing the magnet around their neck.

Bring your kitten home in a cat basket or sturdy cardboard pet carrier. Let her explore the house thoroughly, and don't confront her with other pets until she's had a chance to settle down. Give her plenty of affection, and she'll soon feel part of the family.

WHO WROTE THE OPENING BARS OF THE CAT FUGUE?

Legend credits Pulcinella, pet of 17th-century Italian composer Domenico Scarlatti, who used to idle the time away by wandering up and down his master's harpsichord keyboard, with this feat. Pulcinella had a musical ear, and would sit and listen intently until the vibrations of the notes had completely died away.

One evening, Scarlatti was woken from a snooze by the familiar sound of his cat's perambulations. But this time, instead of the usual random cacophony, Pulcinella seemed to be picking out a melody. Scarlatti grabbed his notebook, transcribed the cat's trail of notes on to staff paper, and used the phrase at the start of *Fugue in G Minor: L499*, popularly known as *The Cat Fugue*.

WHICH CAT LIKES TO LIVE WITH ITS FAMILY?

Most of the 39 feline species much prefer to live alone, and shed no tears when they leave their siblings behind and take off on a solo existence. But the male cheetah loves his brothers, and may sustain a connection with them through life. Lady lions also tend to stick together, in matriarchal prides.

✺ HOW CAN A CAT LOVER WASTE TIME ON A KEYBOARD? ✺

```
     *        ,                    MMM8&&&.   *
                               MMMM88&&&&&    .
    *                        MMMM88&&&&&&&
                            MMM88&&&&&&&&&
                            MMM88&&&&&&&&&
                             'MMM88&&&&&&'
                              'MMM8&&&'    *
           |\___/|
           )     (        .              '
          =\     /=
           )===(        *
          /     \
          |     |
         /       \
         \       /
    _/\_/\_/\__  _/\_/\_/\_/\_/\_/\_/\_/\_
      | | | |(( | | | | | | | | | |
      | | | |)) | | | | | | | | | |
      | | | |(_(| | | | | | | | | |
      | | | | | | | | | | | | | | |
      | | | | | | | | | | | | | | |
```

✺ CHOOSING A CAT ✺

Rescue centers do a great job – but some do it better than others. Even though rescue center owners almost always love cats to bits, if the place is chaotic and not especially clean it's better to walk away. Look for a center that's well-organized and hygienic, and where the cats seem happy and healthy.

Cats that have been abused or lived wild may be adopt to home successfully. A better bet is often a kitten that's been born "accidentally" to an unneutered mother, and has spent its first weeks in the midst of a busy household. A young cat that's been handled by loving humans, and isn't fazed by televisions blaring, doorbells ringing, dogs barking, babies crying or the other sights and sounds of family life, will be well able to cope with a new home and should settle down easily.

❧ UNINTENTIONAL JOURNEYS ❧

Cats love containers. Carrier bags, boxes, drawers, filing trays – you name it, they'll climb into it.

That essential feline curiosity can be dangerous, though, if the container in question is somewhat larger. A ship's container, for instance.

One cat, nicknamed Chairman Meow, even though she was a female, took a nap in a crate of crockery that was being packed in China for shipment to Britain. When she woke, the container had been sealed and loaded. Unlike other creatures, trapped cats are programmed not to panic, and tend to conserve energy if they're in a tight spot, by keeping very still and calm.

This strategy must have done the trick for Chairman Meow because, 26 days later, the crate reached its destination and there she was, weak, but still alive. She must have chewed cardboard and licked condensation to survive. Although thin, she hadn't suffered any permanent harm, and was found a new home in the United Kingdom.

An American cat, Emily, somehow snuck into a container that was destined for Nancy in France. Fortunately, she was wearing a tagged collar, and was reunited with her owners when Continental Airlines flew her – business class – back home.

The port cat at New Plymouth, New Zealand, was owned by Colin Butler, manager of Taranaki tanker terminal, and became known as Colin's Cat or Colin's, for short. She quickly determined that if she mewed loudly enough, people would feed her, and would use this trick to good effect on the off-going shift – and again, half an hour later, with the incoming workers.

But she made a bad move when she begged the second engineer of the tanker Tomawaki for more dinner. He took her on board, meaning to feed her and take her back to land, but after dinner they both fell asleep, and didn't wake up until the tanker had left port.

The upshot – Colin's took an 18-day journey to Korea, where she was promptly popped into a pet carrier and put on the next plane home. The mayor and 50 staff turned out to greet her, but Colin's couldn't cope with stardom. She decided to lead a more sedate life from then on, wandering down to the wharf only occasionally, to chase a seagull, or catch a whiff of the sea breeze and remember her oceangoing days.

❧ HOW DO CATS AFFECT THE HARVEST? ❧

A good harvest was crucial to winter survival in ancient times, when cats were sacrificed to protect the crop. Children were warned to stay off the fields, otherwise the enormous ghostly cat that kept watch over the corn would catch them.

If a man's scythe slipped, a cat was called to lick the wound. Otherwise, the crop was in danger.

A bucolic French festival saw the local cats garlanded with ribbons and fresh flowers when the harvest began. The Japanese stood magical statues of cats in their granaries to keep rats at bay.

❧ WHO'S AFRAID OF PRETTY POLLY? ❧

In his book *La Menagerie Intime*, the French poet, novelist and inordinate cat lover Théophile Gautier (1811–1872) wrote at length about his cats.

Top of the list was Childebrand, whose symmetrical markings made it look as though he was wearing stripy hose. Then came Don-Pierrot-de-Navarre, who liked to sleep draped over the back of Gautier's bedstead. Another cat was Madame Théophile, a ginger cat with pink paws, white chest and blue eyes. She shared the author's meals and "frequently intercepted a choice morsel on its way from my plate to my mouth." And she had a very alarming experience, with a green parrot.

One day, wrote Gautier, a friend who was going away for a short time, brought me his parrot, to be taken care of during his absence. The bird, finding itself in a strange place, climbed up to the top of its perch by the aid of its beak, and rolled its eyes (as yellow as the nails in my armchair) in a rather frightened manner, also moving the white membranes that formed its eyelids.

Madame Théophile had never seen a parrot, and she regarded the creature with manifest surprise. While remaining as motionless as a cat mummy from Egypt in its swathing bands, she fixed her eyes upon the bird with a look of profound meditation, summoning up all the notions of natural history that she had picked up in the yard, in the garden, and on the roof. The shadow of her thoughts passed over her changing eyes, and we could plainly read in them the conclusion to which her scrutiny led: "Decidedly, this is a green chicken."

This result attained, the next proceeding of Madame Théophile was to jump off the table from which she had made her observations, and lay herself flat on the ground in a corner of the room, exactly in the attitude of the panther in Gérôme's picture watching the gazelles as they come down to drink at a lake. The parrot followed the movements of the cat with feverish anxiety. It ruffled its feathers, rattled its chain, lifted one of its feet and shook the claws, and rubbed its beak against the edge of its trough. Instinct told it that the cat was an enemy and meant mischief.

The cat's eyes were now fixed upon the bird with fascinating intensity, and they said in perfectly intelligible language, which the poor parrot distinctly understood, "This chicken ought to be good to eat, although it is green." We watched the scene with great interest, ready to intervene if needed. Madame Théophile was creeping nearer and nearer almost imperceptibly. Her pink nose quivered, her eyes were half-closed, her contractile claws moved in and out of their velvet sheaths, slight thrills of pleasure ran along her backbone at the idea of the meal she was about to enjoy. Such novel and exotic food excited her appetite.

All in an instant her back took the shape of a bent bow, and with a vigorous and elastic bound she sprang upon the perch. The parrot, seeing its danger, said in a bass voice, grave and deep, "*As tu déjeuné, Jacquot?*" – (Having lunch, Jacquot?)

This utterance so terrified the cat that she sprang backwards. The blare of a trumpet, the crash and smash of a pile of plates flung to the ground, a pistol shot fired off at her ear, could not have frightened her more thoroughly. All her ornithological ideas were overthrown.

"*Et de quoi? Du rôti du roi?*" (What are you having? The king's roast?) continued the parrot. Then might we, the observers, read in the physiognomy of Madame Théophile, "This is not a bird, it is a gentleman. It talks."

> "*Quand j'ai bu du vin clairet,*
> (When I drink light red wine)
> *Tout tourne, tout tourne en cabaret,*"
> (Everything becomes a cabaret.)

shrieked the parrot in a deafening voice, for it had perceived that its best means of defense was the terror aroused by its speech.

The cat cast a glance at me which was full of questioning. But as my response was not satisfactory, she promptly hid herself under the bed, and from that refuge she could not be induced to stir during the whole of the day. The next day, Madame Théophile plucked up her courage and made another attempt, which was similarly repulsed. From that moment she gave it up, accepting the bird as a variety of man.

WHAT'S SPECIAL ABOUT MR. PAWS, SNOWY, TIBBLES AND TUFTY?

They belong to Mrs. Figg, a character in J.K. Rowling's *Harry Potter* series, who surreptitiously watches over the young Harry while he lives with the Dursleys. She lives two streets away from Privet Drive, and occasionally looks after Harry, who hates it because her house smells of cabbage and Mrs. Figg makes him look at photographs of her cats.

But Mrs. Figg isn't just a mad old lady. She's a squib, someone who was born into a wizarding family, but doesn't have any powers. Squibs can choose to live in the magical world, or else they live discreetly alongside Muggles (humans). At a glance, Mrs. Figg may live an ordinary life, but in fact she does a roaring trade in cross-breeding cats with Kneazles, their magical variant.

✿ HOW TO PREVENT LITTLE MISHAPS ✿

Tiger Tom backs up to the sofa, a bag of trash or, worst of all, your leg and lets loose with a pulsating spray of pee.

Of course, he's only obeying his natural instinct to mark the sofa, the trash bag or, worst of all, your now dampened leg as belonging to him and no other cat, but that's scant comfort when the reek hits your nostrils.

How to stop him? Neutering is the best route, if the scoundrel belongs to you. If it's someone else's pestilential pet that's doing the marking, then fear is the key. Try drenching him with a well-aimed water pistol, or making a loud clatter before he's too comfortably positioned.

Most spraying goes on out of doors, as part of cats' boundary marking. A cat who starts spraying indoors might be feeling frightened or unsettled. Any change, from redecorating or shifting the furniture, to bringing home a new baby or, horror of horrors, another cat, could trigger a bout of indoor spraying.

When things do change in your home, it's wise to introduce your cat to the changes gradually. Spend time with him in a newly-decorated room, talking to him softly and letting him sniff around until he begins to realize that this is still home. If the worst happens, wash the sprayed area thoroughly with biological detergent, then rub over with surgical spirit and don't let your cat near until the whole place is dry.

Cats don't like spraying near their food, so as another deterrent you could place little tubs of dry cat food near the most likely spraying spots. You can also put down trays of gravel or pine cones or try sheets of foil, which might be enough to deter the would-be sprayer.

TRAINING DAY

Our son brought home a stray cat. We took him
in, but he soon adopted the back of my favorite
armchair as a scratching post. "It's OK," said my son.
"I'll train him." And every day for the next week, he plucked
the cat off the chair and dumped him outside the back door
every time he scratched, to show him he was in the wrong.
The cat soon got the message. And for the next 15 years,
whenever he wanted to go out, he just scratched
the back of my favorite chair.

❧ HOW TO GIVE A CAT A PILL ❧

1. Pick cat up and rest it in the crook of your left arm as if you were holding a baby. Place right index finger and thumb on either side of cat's mouth and apply gentle pressure to its cheeks while holding pill in right hand. As cat opens jaws pop pill into mouth. Allow cat to close mouth and swallow.

2. Retrieve pill from arm of chair and cat from behind sofa. Place cat on left arm and repeat process.

3. Retrieve cat from bedroom, and discard wet pill.

4. Remove fresh pill from bottle, hold cat in left arm clutching rear paws tightly with left hand. Squeeze jaws open and push pill to back of mouth with right index finger. Hold mouth shut to a count of 10.

5. Retrieve pill from DVD player and cat from the china cabinet. Call partner in from garage.

6. Squat on floor with cat lodged firmly between knees, hold front and rear paws. Ignore rumbling yowls from cat. Get partner to hold cat's head firmly with one hand while forcing plastic ruler into its mouth. Slide pill down ruler and massage cat's throat actively.

7. Retrieve cat from window curtains. Fetch another pill from bottle. Make note to buy new ruler and repair curtains. Carefully brush shattered porcelain and glass ornaments from fireplace and collect for gluing later.

8. Wrap cat in bath towel and get partner to sit on cat with its head just visible from beneath his left thigh. Place pill in one end of drinking straw, open the cat's mouth with pen and blow hard down other end of straw.

9. Check medicine container label to ensure pill not poisonous to humans. Drink a beer to remove taste from mouth. Apply bandage to partner's arm and remove bloodstains from carpet with warm water.

10. Retrieve cat from neighbor's greenhouse. Fetch another pill. Open another can of beer. Place cat in cupboard and close door on neck to leave only head showing. Force mouth open with soup spoon. Catapult pill down throat with rubber band.

11. Fetch screwdriver from utility room and replace cupboard door on hinges. Drink the beer you opened in Step 10. Take bottle of malt whisky from sideboard. Pour large one; drink. Apply cold compress to face and check diary for date of last tetanus booster. Apply malt whisky compress to face to disinfect. Swig back another large one. Throw T-shirt away and fetch clean one from bedroom.

12. Phone the fire department to retrieve cat from tree across road. Sympathize with neighbor who crashed into a post while swerving to avoid cat. Take last pill from container.

13. Find heavy-duty pruning gloves in greenhouse, tie cat's front legs to rear legs with baling twine and bind tightly to leg of dining room table. Press pill into mouth followed by large piece of steak. Hold head vertically (pointing upwards) and pour two liters of water down throat to wash pill down.

14. Consume remainder of malt whisky. Get partner to drive you to emergency room, sit patiently while doctor stitches fingers and arm, and removes pill fragments from right eye. Call in at furniture emporium on way home to order new dining room table.

15. Send cat back to cat rescue center and buy a hamster.

HOW TO GIVE A DOG A PILL

1. Wrap it in a piece of cheese and give to dog.

❧ J.G. BALLARD ON CATS ❧

Novelist, essayist and short-story writer James Graham Ballard was born in
Shanghai, China, in 1930. Interned by the Japanese during the Second World War,
his family returned to Britain in 1946. He is author of the best-selling novel *Empire
of the Sun*, as well as *Cocaine Nights*, *Super-Cannes*, *Millennium People* and *Kingdom
Come*. Here, he describes the room where he writes:

> *A Paolozzi screen-print is resting against the door,*
> *which now serves as a cat barrier during the summer*
> *months. My neighbor's cats are enormously affectionate,*
> *and in the summer leap up on to my desk and then*
> *churn up all my papers into a huge whirlwind. They*
> *are my fiercest critics.*

❧ WHO WAS THE ONLY CAT TO WIN THE "ANIMALS' VICTORIA CROSS"? ❧

SIMON, ALSO KNOWN AS BLACKIE, OF HMS AMETHYST

The only cat ever to win the People's Dispensary for Sick Animals Maria Dickin
medal for bravery, Simon was smuggled on board the *HMS Amethyst* in Hong Kong
in 1948, under the uniform of seaman George Hickinbottom.

Fortunately the captain was a cat lover, and he allowed Simon to stay, as long as
there was no trouble. Simon, known to the crew as Blackie, endeared himself to all
on board. He was an efficient ratter, enjoyed a snooze in the captain's upturned cap,
and would come when whistled for. His party trick was fishing ice cubes out of a jug
of water.

Time moved on. Another captain took command, and in April 1949, *Amethyst*
was ordered to relieve *HMS Consort* of the task of guarding the British Embassy
in Nanking, China. She set sail up Yangtze River, but after only 100 miles the
Communists opened fire on her.

The bridge and wheelhouse took direct hits, and *Amethyst* ran aground. The
captain, medical officer and more than 20 of the crew were dead or badly injured.
Simon, who was probably sleeping in the captain's cabin when the ship was shelled,
had his whiskers burned off, and suffered shrapnel wounds to his leg and back. He

survived, though, and as he gradually returned to health, went back to catching rats. He also used to visit the injured seamen in the sick bay, where his comforting, purring presence helped to soothe the traumatized men.

Eventually, after more adventures, *Amethyst* made it back to Hong Kong in summer 1949. The crew – and Simon – became heroes after their experiences in the Yangtze, Simon was awarded the Dickin Medal, or "Animals' VC" for his gallantry. His citation read:

> *Served on HMS Amethyst during the Yangtze Incident, disposing of many rats though wounded by shell blast. Throughout the incident his behavior was of the highest order, although the blast was capable of making a hole over a foot in diameter in a steel plate.*

The story has a sad ending because Simon didn't live to receive his medal in person. When the *Amethyst* reached Britain, Simon went into quarantine. He became ill with a virus and, perhaps because his wounds had weakened him, died peacefully.

He was buried in the PDSA's animal cemetery at Ilford, and these words were carved on his memorial stone:

IN
MEMORY OF
"SIMON"
SERVED IN
H.M.S. AMETHYST
MAY 1948 – SEPTEMBER 1949
AWARDED DICKIN MEDAL
AUGUST 1949
DIED 28TH NOVEMBER 1949.
THROUGHOUT THE YANGTZE INCIDENT
HIS BEHAVIOR WAS OF THE HIGHEST ORDER

ODE ON THE DEATH OF A FAVORITE CAT, DROWNED IN A TUB OF GOLD FISHES

'Twas on a lofty vase's side,
Where China's gayest art had dyed
The azure flowers, that blow;
Demurest of the tabby kind,
The pensive Selima reclined,
Gazed on the lake below.

Her conscious tail her joy declared;
The fair round face, the snowy beard,
The velvet of her paws,
Her coat, that with the tortoise vies,
Her ears of jet, and emerald eyes,
She saw; and purred applause.

Still had she gazed; but 'midst the tide
Two angel forms were seen to glide,
The genii of the stream:
Their scaly armor's Tyrian hue
Through richest purple to the view
Betrayed a golden gleam.

The hapless nymph with wonder saw:
A whisker first and then a claw,
With many an ardent wish,
She stretched in vain to reach the prize.
What female heart can gold despise?
What cat's averse to fish?

Presumptuous maid! with looks intent
Again she stretched, again she bent,
Nor knew the gulf between.
(Malignant Fate sat by, and smiled)

The slippery verge her feet beguiled,
She tumbled headlong in.

Eight times emerging from the flood
She mewed to every watery god,
Some speedy aid to send.
No dolphin came, no Nereid stirred;
Nor cruel Tom, nor Susan heard.
A favorite has no friend!

From hence, ye beauties, undeceived,
Know, one false step is ne'er retrieved,
And be with caution bold.
Not all that tempts your wandering eyes
And heedless hearts, is lawful prize;
Nor all that glisters gold.
Thomas Gray (1716–1771)

∙∘ WHERE THERE IS A WILL … ∘∙

The celebrated 17th-century harpist Mademoiselle Dupuy believed that her musical talents were influenced by her cat, who always sat beside her when she performed, and could show by its behavior whether she was playing well, or badly.

When she died, she stipulated that no hunchbacks, cripples, or blind persons should attend her funeral and left two houses and a handsome income to her cat. Her anonymous feline, Svengali, never received the bequest because Dupuy's relatives went to court to have the will overturned.

∙∘ WHICH "SUNDAY COMPOSER" LET HIS CATS WANDER ON THE DINNER TABLE? ∘∙

Alexander Borodin (1833-1887), scientist and amateur composer, was a tall, dark-eyed stunner of a man with a brilliant and fascinating personality, and a romantic background. He was the illegitimate son of a Georgian prince, who registered him as the son of a serf.

Despite his humble beginnings, the boy Borodin was given a good education. But although he was a keen pianist, he eventually focused on science and became professor of organic chemistry at the Academy of Medicine in St. Petersburg, Russia. He said that music was "a relaxation, a pastime which distracts me from my principal business, my professorship." But the truth was that he adored composition, and often felt hopelessly torn between the lab and the piano. He managed to produce several symphonies and some chamber works, and his best-known piece is the "Polotsvian Dances" from the opera *Prince Igor*, from which the theme for the song "Stranger in Paradise" was taken.

Borodin was a compassionate man, and could never turn away a homeless human – or cat. Consequently, his home was overrun with felines who soon got the upper paw over their soft-hearted patron. Fellow composer Nikolai Rimsky-Korsakov described dinner with Borodin. Throughout the meal, numerous feline lodgers, including Fisher, named for his hobby of catching fish through holes in the ice of the frozen river, "marched back and forth on the table, thrusting their noses into the plates or leaping on the backs of the guests." One cat made an assault on Rimsky-Korsakov's main course, while another draped himself around Borodin's shoulders: " 'Look here, sir, this is too much!' cried Borodin, but the cat never moved."

❧ WHAT'S NEAT ABOUT MEAT? ❧

It contains the amino acid taurine, a compulsory part of the feline diet. Without enough of this vital nutrient, your cat could develop major heart or eye damage. Let her eat meat!

> *Wherever the cat of the house be black,*
> *the lasses of lovers will have no lack.*
> Old saying

❧ WHO MAKES IT LUCKY 14? ❧

Kaspar, a three-foot-high wooden sculpted black cat, made by Basil Ionides, comes to dinner whenever a party of 13 book a table in the Savoy Grill, London.

He was made on commission in the 1920s, so that superstitious guests needn't live in fear. The story began in 1898, with Woolf Joel, who hosted a dinner for 13, and unwisely mocked the old saying that the first person to rise from a table of 13 would meet an untimely end. He was shot after he returned to Johannesburg, South Africa, and the Savoy Hotel decided it was time to play it safe.

For a few years, a staff member would sit in on "unlucky" dinner gatherings, but then Kaspar was commissioned. Just in case those unlucky demons should get the wrong idea, a place is set for him, he's seated at the table and, far from gazing implacably at a plate of wooden cat food, he's served every course. Kaspar was stolen more than once by high-spirited military officers during the Second World War, but was always returned unharmed.

"I have studied many philosophers and many cats.
The wisdom of cats is infinitely superior."

Hippolyte Taine, 19th-century
French critic and historian

⁖ INTROVERT OR EXTROVERT? ⁖

Like people, cats have distinct personality types. Is your cat warm and friendly, or
more of a shrinking violet?

FRIENDLY AND SOCIABLE

Ideal pets, these cats like lots of company from people and other cats, and have a
relaxed attitude to life. They like nothing better than plenty of fuss and stroking,
and are happiest when their people are within eyeshot. You cannot go wrong
with this type, but if you're out a lot during the day consider providing a feline
companion.

NERVY OR EXCITABLE

Cats in this group really do walk by themselves. They aren't that keen on people or
other cats, although they may put up with the attentions of their nearest humans.
You'll seldom find them snuggled on a lap, and if their owner tries hard to seduce
them they'll simply retreat even further. Don't overdo your efforts to persuade this
cat to love you. Tempt him with food and warmth, but don't smother him with
affection when he does seek you out.

⁖ COULD A CAT WIN A MARATHON? ⁖

It's not often that you'll see a cat in a hurry, unless there's an enticing bird – or bowl
– that's worth breaking into a trot for.

Pikachu's default movement is the gentle amble, and this is not because cats are
born idle. They have the ability to spring into action in the blink of an eye, and can
sprint, leap and bound with incredible balletic grace and strength. But, the one thing
they don't have is staying power.

That sleek feline body is well-muscled, but many of the muscle cells are designed
for instant speed, not stamina. Master of the quick sprint, Pikachu would never
make a marathon runner, because all that fabulous energy is quickly consumed.

That's why cats will play energetically for a few minutes, then suddenly stop and stroll away. The poor dears are simply worn out.

At a gentle jog-trot, a cat's body uses up more energy than that of a dog of equal size and weight. Where dogs are designed for the prolonged chase, cats are put together for the quick-response followed by a recovery period. After a dash of less than a minute, a cat is in so much of a sweat that it has to stop and pant.

✺ COULD A CAT BE MUM TO A BIRD? ✺

Not often, it's true, but there was a Brazilian cat who found a fledgling that had toppled out of the nest, and raised it as if it were a kitten. The unlikely duo would eat together from the same bowl, the bird not seeming to mind that it was eating a peculiarly meaty diet.

✺ HOW DO YOU SAY CAT IN …? ✺

Off to exotic climes? Here's how to attract a cat in some
far-flung locations:

African dialect – *katsi*
Arabic – *biss/hirrah/qitah/quttah*
Cherokee – *wesa*
Chinese – *mao*
Farsi – *gorbe*
Filipino – *pusa*
Fula (New Guinea) – *gnari*
Gujarati – *biladi*
Hawaiian – *popoki*
Hindi – *billy*
Japanese – *neko*
Korean – *ko-yang-ee*
Indonesian – *kucing*
Russian – *kot/koshka* (m/f)
Swahili – *paka*
Thai – *meo*

❀❀ POLYDACTYL PETS – OWNERS TALKING ❀❀

Imagine a cat with an extra toe or two, and that's "polydactyl."
Read more about them on page 54.

*Our last poly kitten could pick up pens and pencils, get
the lid off her food container and open cupboard doors.
We put elastic bands on the door knobs to add tension.
But that didn't stop him; it just slowed him down.*

*My Maine Coon Ernie was named after Ernest
Hemingway, and has a total of 27 toes – eight on one
front paw, seven on the other and six on each back foot.
He looked so sweet when he was a kitten, with these
huge baseball mitt paws on his little stick legs!*

*I once owned an Abyssinian cat that could open doors
and pick things up. He also could jump up on the counter,*

open the cabinet door, get a box of cat treats and start
eating them – without us hearing him. Very stealthy!

Mittens would pick up dry food in a bowl, dip it in
water and eat it off the center of his paw. I'm sure
he'd like to hitchhike, and I'm just waiting for him to
give me the thumbs-up because he likes his dinner!

My Freddie can get on the countertop and open
the cabinet above it. He opens the glass doors, then
closes them behind him and sits in there with the
wine glasses, looking out at us through the door. His
giant "thumbs" work just like ours. He can pick up
coins from the floor, take pens out of the cup and
even hold a toothbrush. He's ginger with long white
hair on his paws. They look just like fluffy mittens.

I own Aaron, who has 12 toes in total, and I also
had his dad, Sam, who was totally black and had
13 toes. The vet said he was unlucky.

CATS AND COMPUTERS 1
Surely humans can type on one half of the
keyboard while I use the other
side as a headrest?

IS IT TRUE THAT WOMEN PREFER CATS?

Dogs may still claim to be "man's best friend," but a woman's best friend, after her
cellphone, is now definitely her feline. Cats are pet of choice for 4.7 million U.K.
women – almost four times the number of male owners. In fact, dog days could be
seriously numbered, since cats now outnumber dogs as favorite U.K. pets by a cool
1.5 million.

☙ HOW DO CATS THINK? ❧

Ah, how strange are the ways in which animal behaviorists amuse themselves.

For instance, experimenters have conducted a whole range of tests to try and find out just what does go on in Tom Kitten's furry head. Can he, for instance, behave like a human child, and find a hidden object by using his powers of learning and deduction? So, if a friendly researcher shows T.K. his favorite toy, puts it into a container while he watches, takes the container behind a screen and removes the toy, then returns to show our bemused feline the empty container does he – or does he not – know what the heck is going on? Can he figure out that the toy is now behind the screen?

And the answer is "Yes," but it's a qualified "Yes." Cats, it appears, do create a mental map of their surroundings, and can deduce what has happened to a hidden toy. But if they become at all confused – and who wouldn't, with this kind of scientific nonsense going on – they will use themselves as a central point, and look for things where last seen in relation to themselves, rather than following logic as a human would. Which explains why, if Tom is sitting at point A, with his food bowl at point B, and you move his bowl to point C as he watches, he'll still go to point B and look puzzled.

"Cats are a mysterious folk.
There is more passing
in their minds than we are aware of."
Sir Walter Scott (1771–1832)
prolific Scottish novelist and poet

☙ ARE FOXES A DANGER TO CATS? ❧

If you live in an urban area and look out of your window as dusk falls, you're quite likely to see a fox trot confidently across your yard. As more and more foxes move into towns and cities, the resident cats have to learn to share their territory with Mr. Tod and his family.

It's often said that if a cat and fox meet, the cat will lose, but the truth is that the fox is more likely to turn tail and run. Foxes have no interest in attacking cats, and won't hang around to make friends.

CAN YOU IMAGINE ...
JOHN LENNON'S CATS?

The Beatle with the "Who cares?" attitude had a super-soft spot for cats.
Here are some of the fab felines that shared his life.

TICH

A ginger half-Persian was one of John's childhood pets and died during the years he
was studying at Liverpool College of Art.

TIM

Another fluffy cat. John found him wandering in the street and took him home,
where he became a special favorite.

SAM

Third of John's childhood pets. The Beatle-to-be would bike over to the fish shop
every day, to buy his pets a nice piece of hake. Even in his 20s, John would phone
home from tours to see if the cats were missing him.

MIMI

Although John frequently saw his mother through his childhood, he spent most
of it living with her sister, his Auntie Mimi, another cat lover. In the 1960s, when
Lennon, his first wife, Cynthia, and their son, Julian, lived in Weybridge, their first
cat, a tabby, was named after Mimi. They had another tabby named Babaghi, and
eventually owned about 10 cats.

MAJOR AND MINOR

A pair of cats, one black, one white, belonging to May Pang. Lennon lived with her
for 18 months between 1973 and 1975, but was eventually reconciled with Yoko
Ono, whom he had married in 1969.

PEPPER

John Lennon, Yoko Ono and their son, Sean, shared their lives with various cats.
The black Pepper was twinned with a white cat, Salt. Misha, Sasha and Charo also
resided at the Dakota, the grand 19th-century New York apartment block outside
which Lennon was murdered in 1980.

❧ MILK FOR THE CAT ❧

When the tea is brought at five o'clock,
And all the neat curtains are drawn with care,
The little black cat with bright green eyes
Is suddenly purring there.

At first she pretends, having nothing to do,
She has come in merely to blink by the grate,
But, though tea may be late or the milk may be sour,
She is never late.

And presently her agate eyes
Take a soft large milky haze,

And her independent casual glance
Becomes a stiff, hard gaze.
Then she stamps her claws or lifts her ears,
Or twists her tail and begins to stir,
Till suddenly all her lithe body becomes
One breathing, trembling purr.

The children eat and wriggle and laugh;
The two old ladies stroke their silk:
But the cat is grown small and thin with desire,
Transformed to a creeping lust for milk.

The white saucer like some full moon descends
At last from the clouds of the table above;
She sighs and dreams and thrills and glows,
Transfigured with love.

She nestles over the shining rim,
Buries her chin in the creamy sea;
Her tail hangs loose; each drowsy paw
Is doubled under each bending knee.

A long, dim ecstasy holds her life;
Her world is an infinite shapeless white,
Till her tongue has curled the last holy drop,
Then she sinks back into the night,

Draws and dips her body to heap
Her sleepy nerves in the great armchair,
Lies defeated and buried deep
Three or four hours unconscious there.

Harold Monro (1879–1932)

Monro was an English poet who founded the *Poetry Review*. This is probably his best
known poem, and its beauty lies in the detail and the empathetic observation.

✿ WHY IS A CAT CALLED A CAT? ✿

Every domestic cat in the world today – and there are hundreds of millions of them – is thought to be descended from the North African wild cat *Felis libyca*.

Biological evidence points to Libya in North Africa as the birthplace of the cat. The nomadic Berbers, who inhabited the northwestern Libyan plains before Arab cultures arrived there, called this beautiful wild creature *kadiska*, and it's thought that the word "cat" and its numerous variations in other languages stems from that first feline soubriquet.

Other North African and Middle Eastern languages have similar words – the ancient Nubians of northern Sudan referred to their pets as *kadis*. To the Syrians, they were *qato* and in Arabic, *quttah*.

When is a cat not a cat? To the ancient Egyptians, who must have been as familiar with the sound of insistent mewing as we are, and chose the onomatopoeic name *miou*. The pet name "pussy" perhaps comes from the name of the early Egyptian cat goddess Bast (pronounced Pasht).

The oldest Latin texts used the word *felis* to identify not just cats, but any mid-sized, brownish carnivorous mammal, including the weasel and the polecat. By A.D. 300, cats had a Latin word all of their own – *cattus* – which then changed into the Old English *catte*.

CATS AND COMPUTERS 2
The screen makes such a
divine backlight to my tail.

✿ WHAT IS THE DICKIN MEDAL? ✿

Maria Dickin (1870–1951) was an intelligent and active woman. She married in 1898, and, although late Victorian married women were expected to give up work and look after their households, Maria needed something more satisfying.

She began doing some social work, but her sheltered upbringing meant she was completely unprepared to witness the kind of suffering and deprivation common among poor people – and animals – at the end of the 19th century.

She saw maimed and diseased animals scavenging a living from the gutter, an unbearable sight. "The suffering and misery of these poor, uncared-for creatures in our overcrowded areas was a revelation to me. I had no idea it existed, and it made me indescribably miserable."

The problem haunted her, and she made up her mind to do something to help. Finding funds and premises was a struggle, but at last, in 1917, the first-ever People's Dispensary for Sick Animals was opened in London. A notice above the door read:

Bring your sick animals.
Do not let them suffer.
All animals treated.
All treatment free.

Crowds flocked, the police were called in, and the clinic had to move to larger premises to treat more than 100 animals a day.

Maria Dickin went on to establish PDSA clinics throughout the United Kingdom and abroad. A remarkable woman, she devoted her life to helping creatures who can't speak up for themselves have lives that are free of pain and suffering.

During the Second World War, Maria realized that animals could be just as brave as humans, and she instituted the Dickin Medal. Recognized as the animals' Victoria Cross, the medal is still given to animals who display conspicuous gallantry and devotion to duty while serving or associated with any branch of the armed forces or civil defense units. It has been awarded to one cat, Simon of *HMS Amethyst* (see page 34).

WHEN DOES LILAC TURN INTO LAVENDER?

When it's being described by one of the cat breed associations, who like to confuse by giving the same genetic color a different name, depending on which breed they're discussing …

- Lilac (a pale, creamy shade) may be called Lavender.
- Black can be Ebony.
- An Oriental Shorthair whose coat is genetically Chocolate can masquerade as Havana or Chestnut.
- Ginger or Marmalade cats' self-colors, or markings, are called Red in the pedigree cat world, or occasionally Auburn.
- Tortoiseshell and white cats are sometimes known as Calicos in North America.

❖ HOW CAN CATS FIGHT SPAM? ❖

How do you tell the difference between a dog and a cat? Humans know instantly which is which, but it's hard to define exactly what makes a dog so doggy, and a cat so instantly recognizable as feline.

It's that unquantifiable difference that computer experts have latched on to, as a way to outfox malicious computer hackers who use automated systems to set up thousands of free e-mail accounts in order to bombard the rest of us with junk mail.

One current method of deterring spammers uses a combination of letters displayed on a distorted grid, known to geeks as a CAPTCHA – Completely Automated Public Turing Test to tell Computers and Humans Apart. Anyone setting up an online email account has to type in a Captcha code in order to prove they're a real person.

Trouble is, often humans can't make sense of Captcha codes – and computers can. So researchers have been looking for images that humans can easily tell apart, but where the differences are subtle enough to confuse a robot.

And they've found an answer in (stand by for another acronym) ASIRRA (Animal Species Image Recognition for Restricting Access), which asks users to identify all the cats or dogs from a grid of several tiny pictures. For a human it's a piece of cake,

but for all computers can tell something that's furry, four-legged and has two ears, eyes and a nose could just as easily be dog, a cat – or a wombat.

Asirra draws on a bank of more than two million images of cats and dogs from American animal shelters – and you can even click on the "adopt me" button, if one of those appealing faces melts your heart. If the system becomes widely used, it could reduce spam, and make Internet browsing a warmer experience.

"My cat has taken to mulled port and rum punch. Poor old dear! He is all the better for it. Dr. W. B. Richardson says that the lower animals always refuse alcoholic drinks, and gives that as a reason why humans should do so, too."

In a December 24, 1879, letter to Samuel Butler, his correspondent reveals her pet's drinking habits.

WHAT DO CATS AND CHICKENS HAVE IN COMMON?

You mean you didn't know?

Neither of them can taste sweetness.

Taste is transmitted through genetic receptors on the tongue that signal to the brain – mmm, sugar or, bleck, lemon. Cats – and chickens – missed out on a particular gene that combines with another to detect sweetness, and so neither of them will ever experience the delights of a Kit-Kat or a Cadbury Crème Egg.

Although cats can distinguish between salty, bitter and sour flavors, they are the only mammal that can't detect sweetness, possibly because, as meat eaters, they don't need the ability to spot carbohydrates – an important food source for plant-eating species (and humans).

Don't feel too sorry for your deprived pet. Many cats already suffer from bad teeth and although a few will scoff at ice cream, marshmallows and cotton candy, their digestive systems really can't cope with sugar, and they risk developing diabetes if they have too much.

Better eat all the chocolate yourself.

HOW GOOD IS A CAT'S MEMORY?

A cat's recall is better than a monkey's or an orangutang's, and much better than a dog's, according to tests done by the University of Michigan. Where a dog's memory doesn't last more than five minutes, a cat's can be retained for up to 16 hours.

What's a cat's favorite color?
Purrrrrrple!

HOW DO YOU ENTERTAIN AN INDOOR CAT?

Apartment-dwellers all over the world keep cats, whose only contact with the delights of the outdoors is a patio or balcony, if they're lucky. For creatures that are born hunters, a life spent indoors could be somewhat dull. So how can you amuse and stimulate your indoor cat, and give her energies an outlet?

HIDE-AND-SEEK

Play hide-and-seek games. Imitate the way prey appears, then hides, by secreting a toy, piece of string or even your hand under a blanket or newspaper, then popping it out again. Cats will watch fascinated, and try to pounce, so vary the speed of your movement (and watch your hand!).

INSECT AND BIRD SIMULATIONS

Give your cat toys that simulate the flutterings of insects or birds. A garden cane with a length of string attached, and a bunch of feathers tied to the end is a lovely toy. Flip and flick the cane and your cat will run and jump, trying to catch the string.

CATNIP TOYS

Any toy stuffed with catnip is guaranteed to please your cat, but renew these toys regularly, as the delicious scent fades over time.

NOISY AND SPARKLY TOYS

Toys that make a noise or sparkle, such as little woven plastic balls with bells inside, are particularly fascinating to cats, who love to bat them around the floor. Buy lots, as they always end up jammed under the sofa.

MOVING LIGHTS

A moving light is an intriguing plaything. On a sunny day, idle away a happy 10 minutes as kitty tries to catch the bright reflection of the sun on your watch face. Direct the light up the walls or along the floor, and the chase begins.

SCRUNCHED-UP PAPER

Tiny balls of scrunched-up paper are another feline favorite. Your cat will learn to recognize the sound and come running at the rustle of a candy wrapper. Roll the paper into a ball and flip it into the air for your cat to catch.

DARK HIDE-OUTS

Give your cat a dark place to hide out. A cardboard box, or fabric or paper shopping bag (not plastic because of the danger of suffocation) are perfect spots for puss to hide in.

SPREAD NEWSPAPERS

A newspaper spread on the floor is a fantastic cat's playground (better read it first), and they can spend ages ripping, pouncing and rustling.

HIGH SPOTS

Provide a high spot where your cat can roost, and watch the goings on below her. The top of a wardrobe or fridge is good, or you can buy or make a cat-shelf, covered in carpet for ripping. If you have space, fix one or more of these high on the wall so puss can leap between them, dangle from them and generally have all the fun that trees and fences would provide out of doors.

SCRATCHING POSTS

A scratching post is not only necessary to save your furniture (if you can persuade your cat to use it), but also useful for climbing up. Attach a bar at right angles to it, with a dangling ping-pong ball, so your cat can give it a passing swipe when the mood takes her.

SMALL OBJECTS

Any small object left invitingly near the edge of a surface – a coin on a table, a shiny pen – is an open invitation to your cat to poke and prod it until it drops to the floor. Whereupon she can stroll nonchalantly away.

⚫ HEMINGWAY'S CATS ⚫

Where can you come face to face with Emily Dickinson, Simone de Beauvoir, Spencer Tracy, Pablo Picasso, Audrey Hepburn, Joan Crawford, Trevor Howard, Charlie Chaplin, Sophia Loren?

In Madame Tussaud's? At the Museum of Film and Cinema? No, these are the names of some of the 60-plus cats that live in the Ernest Hemingway Museum on Florida's Key West.

Hemingway was always crazy about cats. When he lived near Havana, he shared his house with 30 or more cats, and tried to create a new breed by mating Angoras with the local Cuban cats. He moved to Key West in 1929, and the first book that he wrote there was *A Farewell to Arms.* His house is now a museum and home to a colony of cats.

The Hemingway cats aren't of any particular shape, size, color or breed – but about half of them share one strange characteristic. They are polydactyl cats – they have extra toes. Usually, cats have five toes at the front and four on their back paws. Polydactyls can have one or more extra toes at the front and sometimes also on the back. It looks almost as though they're wearing mittens or boxing gloves, because the extra toe looks like a thumb. Cats with opposed thumbs often become very dexterous with them, and can manipulate objects in a way that's scarily human.

Hemingway, so the story goes, was given a polydactyl cat by a ship's captain, and the cats that live in the museum grounds are descended from that one cat. Key West is so small that most of the cats on the island are probably inter-related.

Sailors have always thought polydactyl cats are lucky, and that may be why they're still so common on the East Coast of America. They arrived on ships and formed colonies. Boston in particular still has a large number of polydactyl cats, known as "mitten cats", "thumb cats," "six-finger cats," or "Hemingway cats."

CATS AND DOGS
Dogs have owners. Cats have staff.

⋅◦ WHICH PEDIGREE BREED IS BEST? ◦⋅

Pedigree cats are gorgeous, but it's best not to let yourself be seduced without doing some homework. All breeds have their own characteristics, so find out more before you go kitten-shopping.

- Have a fun day at a cat show, talk to breeders and owners, look at the different breeds and see which ones appeal.
- Longhaired Persians and semi-longhaired breeds such as the Birman are adorable bundles of fur, but don't even think about owning one if you can't devote daily time to grooming. You'll need to wield those brushes right from kittenhood, so your fluffy feline learns to accept that grooming is a part of life.
- Fancy a chatty cat, like the Siamese, Burmese or one of the Rexes? Only take this route if you're around a lot, and are happy to spend a fair bit of time in cat interaction. These breeds don't just like to talk, they want fun, activity and stimulating play. That distinctive voice will often be raised, as these breeds are very far from silent, so talk to a few owners or visit them at home to find out for yourself what life sounds like with one of these idiosyncratic cats.
- Mind made up? When you go to the breeder, look at the health of all the cats. If there are felines all over the place and it smells unpleasant, then hygiene is probably poor and the cats may be diseased.
- Be prepared not to come away with a kitten if you have any doubts, or if the kitties look unhealthy. Beware the tug on your emotions if you do encounter a forlorn kitty that looks under the weather, but take it on at your own risk. You could be in for big heartache and even bigger vets' bills. If the breeder can't offer a lively, healthy, kitten it's better to look elsewhere.

❀ WHAT DO CATS HATE MOST? ❀

Vacuum Cleaners

Few cats can keep their cool when the vacuum comes to get them. Use this to your advantage if kitty scratches at your bedroom door and wakes you before you're ready. Leave the cleaner within arm's reach and give a quick whirr when you hear the first scratchings. It shouldn't take more than a couple of days for puss to work out that it's safer to let you emerge in your own good time.

A Break in Their Routine

Stick to a regular pattern of feeding times. Cats are creatures of habit and quickly learn when to expect the next bowl of food. If you feed erratically, you'll create a restless pet that pesters you for food whenever its tummy rumbles.

Strong Smells

Those delicate feline nostrils are set all a-quiver when they meet up with citrus peel, ground pepper or slivers of strongly-scented soap. Scatter these items on no-go areas like flowerbeds and tabletops to deter your cat from visiting.

Replicas of Cats

Cats are very suspicious of life-sized cardboard cats or statues. You can set one up strategically to put a scare into cats that invade your garden or amble in through an open door.

Chicken Wire

The feeling of wire under the paws seems to disturb cats, who tend to keep off parts of the garden where you lay the wire down. Useful for preventing plant eating, tree-trunk scratching or indiscriminate toilet habits.

Loud Noises and Wet Surprises

A sudden jet from a water gun or garden spray is enough to send the most persistent marauder scampering away. You can also put your own cat off the idea of jumping on worktops and shelves by positioning a tin can with some coins inside where it will be easily knocked off with a suitably terrifying clatter.

ANIMALS BELONGING TO THE CAT FAMILY

African golden cat	Geoffroy's cat	Oncilla
Andean mountain cat	Iriomote cat	Onza
Asian golden cat	Jaguar	Pallas's cat
Bay cat	Jaguarundi	Pampas cat
Black-footed cat	Jungle cat	Panther
Bobcat	Kod kod	Puma
Caracal	Lion	Rusty spotted cat
Cheetah	Leopard	Sand cat
Chinese desert cat	Leopard cat	Serval
Clouded leopard	Marbled cat	Snow leopard
Eurasian lynx	Margay	Spanish lynx
Fishing cat	North American lynx	Tiger
Flat-headed cat	Ocelot	Wild cat

WHAT, PRAY, ARE THE BARD'S VIEWS ON CATS?

Whether they turn up as witches' familiars, or objects of loathing or fear, cats seldom get good press from William Shakespeare.

Some men there are love not a gaping pig;
Some, that are mad if they behold a cat;
And others, when the bagpipe sings i' the nose,
Cannot contain their urine: for affection,
Mistress of passion, sways it to the mood
Of what it likes or loathes.

Shylock in *The Merchant of Venice*, Act IV

I could play Ercles rarely, or a part to
tear a cat in, to make all split.

Bottom, in *A Midsummer Night's Dream*, Act I

I could endure any thing before but a cat, and now
he's a cat to me.

Bertram, in *All's Well that Ends Well*, Act IV

They'll take suggestion as a cat laps milk
The Tempest, Act II scene I

The queen, sir, very oft importuned me
To temper poisons for her, still pretending
The satisfaction of her knowledge only
In killing creatures vile, as cats and dogs,
Of no esteem:

Cornelius, in *Cymbeline*, Act IV

Where the place?
Upon the heath,
There to meet with Macbeth.
I come, Greymalkin

Three witches, *Macbeth*, Act I
Greymalkin was a gray cat.

Thrice the brinded cat hath mewed ...
First Witch, *Macbeth*, Act IV

◦◦ IS IT WORTH CHATTING WITH MY CAT? ◦◦

Few cats have much of a vocabulary. They probably recognize their name and other familiar words more by the tone of voice, than the actual sound of the word. If you always call your cat on an upnote, try varying the pitch and see if she still responds. Or ask a stranger to call your cat, without signaling to her to come. Chances are that she won't react to her name when it's called by an unfamiliar voice.

Don't let this stop you talking to your cat, though. Cats are great companions, happy to listen to human ramblings, and really won't mind what you talk about, as long as you stroke them while you do it.

◦◦ THE DEVIL'S DICTIONARY ◦◦

Ambrose Bierce, in his satirical work of 1911, *The Devil's Dictionary*, defines the cat as: "A soft indestructible automaton provided by nature to be kicked when things go wrong in the domestic circle." It still isn't funny, Ambrose.

❧ WHY DO CATS LOOK BEFORE THEY LEAP? ❧

Given a chance, every cat will take the time to size up a vertical jump before springing. Of course, cats can leap and bound effortlessly when they're chasing or playing. It's second nature. But if there's a high fence to scale, or a cabinet top that would make the perfect roost, Molly will first of all judge the distance and flex her muscles, ready for take-off.

She'll have a better chance if she can take off from a firm surface, which gives those powerful hind leg muscles the push-off they need. In a pinch, a cat can cover six times its own length in a single bound – although even the most athletic occasionally under-estimate and fall back, and will give themselves a nonchalant lick to show that they're not remotely embarrassed.

The bones of your cat's rear legs are built like a hinge which, together with the fantastically strong muscles in the hindquarters, means she can drop her belly to the ground to creep lightly and silently, or leap swiftly with equal ease. Her razor-sharp claws and strong teeth equip her to deal the death blow efficiently, like a velvet killing machine. The final touch is in the muscles which let Molly rotate her front paws so that she can grasp prey, or climb a tree trunk.

❧ WHY DOES MY CAT EAT WOOL? ❧

This bizarre behavioral glitch crops up most often in Siamese and Burmese cats, but even common or garden kitties sometimes like to snack on a sweater.

And it's not just wool that gets those feline taste buds tingling. Cats also may relish other fabrics, including synthetics and cotton. Clothing that their owners have worn, towels and bed linen are the most appetizing treats.

When Miss Zoe settles down to a fabric feast, she goes into a strange trance-like state. She might stop if you yell, clap your hands or even give her a water-pistol squirt, but minutes later she's back, tearing off mouthfuls of fabric and giving them a good chew before swallowing them with a satisfied purr.

Strangely enough, although the occasional wool-eater does suffer a blockage, the feline digestive system can usually cope with quantities of shredded textiles, which pass through unimpeded and emerge virtually unchanged at the other end without doing any harm.

More of a threat to fabric-eaters than raging indigestion is a ranting owner. Cats have eaten chunks out of designer dresses and expensive carpets, and the habit is hard to break.

- Some cats eat fabrics only when they're lonely or bored. Tackle the problem by encouraging the cat to go out of doors for more stimulation, so that it's less dependent on human interaction.
- Hide edible fabrics wherever possible.
- Nibbling on a specially laid, foul-tasting fabric bait might put Miss Zoe off, but don't bother with mustard, pepper or curry. She'll just develop a taste for exotic flavors. Try menthol or oil of eucalyptus instead.
- Leave a bowl of dried cat food available all day. Some cats just like to snack, and once real food is offered they'll give up eating scarves for good.
- If you can't beat 'em, follow other owners' tips and add some finely chopped undyed wool to Miss Z's normal food, or give her a towel to chomp on in between mouthfuls of meat.

HOW TO BE A HELPFUL CAT

Leave this open where your cat can read it, and look forward to having some
feline assistance around the house.

DOORS

Do not allow closed doors in any room. To encourage someone to open the door,
stand on hind legs, bat at the door handle or mew helplessly. Once the door is opened,
don't feel you have to use it. After you have commanded a door to the outside to be
opened, stand half-in, half-out and muse on the meaning of life. This is especially
important if it's snowing, raining or windy. In your own time, wander back indoors.

RUGS

If you feel sick, run to a rug. A chair is even better. Expensive Oriental rugs are best,
but shag pile is more fun. When throwing up on the carpet, practice a simultaneous
backward movement. Your aim is to produce a barefoot-sized clump.

BATHROOM

Always accompany guests to the bathroom. There's no need to do anything.
Just staring is enough.

HELPING HUMANS

Get as close to your people as you can when they are actively busy with something.

COOKING

When supervising cooking, sit just behind the cook's left heel where you can't be seen.
This gives you the best chance of being stepped on, then picked up and cuddled.

READING

If you can't drape yourself across the book itself, snuggle in under the chin, between eyes and book. Creep up on the outside of newspapers, then leap. You're human will cry out with delight at seeing you.

KNITTING

Settle yourself where you can reach out and bat the needles from time to time. Remember to catch the wool every time your human gives the ball a fresh tug.

PAPERWORK

Sit or lie on the paper being worked on. When removed, watch mournfully from the side of the table. Once your human is absorbed, roll around on the papers to scatter thoroughly. To be extra helpful, push pens, pencils, and rubber bands off the table edge one at a time, or pick up paper clips and elastic bands in your teeth and toss them around.

Cities, like cats, will reveal themselves at night.
Rupert Brooke, poet, author of the famous First World War poem "*The Soldier*", which begins, "If I should die, think only this of me …"

TEN MONTHS PREGNANT AND STILL NO KITTENS?

That's a fat cat! Other sure signs that Fizzy needs to shed a few pounds:

- You've enlarged the cat flap so much, it needs a garage door opener.
- Mice get caught in her gravitational pull.
- She doesn't stalk off in disgust until the third bowl of food.
- That vast fluffy underbelly keeps your wooden floors well polished.
- She has more chins than lives.
- Always lands on her navel.
- You never need to call the fire department, but there's been an upsurge in broken tree branches.
- Your 10-year-old's best friend sat on her – "You mean that's not a beanbag?"

> What do you get if you cross a
> cat with an elephant?
> A big furry creature that squashes
> you when it jumps on your lap.

❧ WHAT SHOULD I GIVE MY CAT FOR DINNER? ❧

Visit the pet food section of a supermarket, and you'll be overwhelmed by the variety. Dry food, wet food, bags, boxes, cans, pouches and mini-cans – what's best? Or should you be filleting salmon for your darling, or tempting her with morsels of finely-chopped liver?

Here are the pros and cons.

WET FOODS (CANS, MINI-CANS, POUCHES)

- A lot of cats like these best, and the jellied varieties seem to be particular favorites.
- May contain up to 75 percent water, which is enough to give your cat sufficient fluid for the day.
- Once in the bowl, wet food quickly turns stale, and in hot weather it starts to smell rancid, dries out and attracts flies. You can keep half-used cans in the fridge, covered, but it's a kindness to your cat to let the food come back to room temperature for 15 minutes before feeding. Remove uneaten food after 30 minutes.
- They don't give the gums a workout, unlike dry foods.
- Can become expensive, especially if your pet develops a taste for the high-end brands. Single-serving pouches are convenient, but, weight-for-weight, far more costly than large cans.

SEMI-MOIST FOODS

- Similar pros and cons to wet foods, but they contain less water. So if you use them, make sure a drink is always available as well.

DRY FOOD

- Most vets think dry food is best for cats because crunching up the morsels helps to keep tartar at bay and avoid gum disease and dental decay, both of which can lead to other health problems.

- Even if your cat is aged and toothless, she can still eat dry food if she likes it. Cats with bad teeth simply scoop up the crunchies with their tongue and swallow them whole.

- Dry foods contain only 10 percent fluid, so it's essential always to serve them with water alongside.

- Very convenient, dry foods keep for ages even after the box is opened, and don't go bad in the dish.

HOMEMADE FOOD

- If you want to custom-cook for your cat and avoid commercial foods altogether, tap into your vet's nutritional know-how first. Cats have various dietary needs which manufactured foods are designed to cover.

- Fish – cook, flake, and check very carefully for bones before feeding.

- Meat – use lean meat or poultry and add bulk with a little rice, pasta or cat biscuit.

- Raw liver – your cat may adore you for it, but there's potential harm in a diet that regularly includes raw liver. Too much can lead to an excess of vitamin A, which causes listlessness, stiffness and even bone problems. If your cat's a liver-lover, give it just once a week.

- Egg – serve it scrambled, never raw.

HUMAN FOOD

- No, no, no! Cassie is not a person, even if she sometimes seems like a fur-clad kid. Many of the foods we happily consume are unsuitable for cats, and too much can make them fat. Offer a little diced chicken breast or a spoonful of a pot pie now and again if you must, but don't make a regular habit of feeding kitty from the table.

VEGETABLES

- Cats are carnivores. They need a high-protein diet that contains plenty of meat. You can give Cass small quantities of vegetables, either cooked or raw, if she'll eat it, but don't make them the main part of her diet.

- One bit of greenery that all cats appreciate is grass. Cats with outdoor access will take a nibble at the lawn, but if your cat's confined, keep a bowl of grass that she can chomp on. Although eating grass does encourage cats to cough up hairballs, that's not a reason for withholding it. A cat that can't find grass most likely will have a munch on your houseplants instead, which could be toxic.

WHAT DOES IT MEAN IF ...

... a cat washes himself as dusk falls? A friend will visit before it's dark.

... you dream about a black cat at Christmas? There will be illness in the family during the year to come.

... a playful cat appears in a dream? A friend is untrustworthy.

*A cat who wants breakfast
has no snooze button.*
Anon.

HOW DOES MY CAT KNOW WHEN I OPEN THE FRIDGE?

It's rather entertaining to watch Dolly Mixture twitch and twirl her ears. She can flex them together or one at a time – an enviable feat, but one we humans will never be able to emulate, as our deeply inferior ears have only six muscles, where Dolly's are equipped with 30. That's how she can flex and rotate her ears to catch passing sounds.

You know how your cat magically appears, the moment she detects your presence. Cats rely heavily on sound to indicate their surroundings. "Ah-hah – the fridge door? I'll go and ask them to feed me."

The cupped, triangular feline ear is perfectly shaped so that it can pinpoint and magnify sounds. Inside the skull, echo-chambers turn up the volume, helping their owner to home in on even the tiniest, most distant noise.

Dogs are renowned for detecting super high-pitched sounds, but a cat's ears are even better (of course!). Dolly M can detect pitches an octave-and-a-half higher than the human ear can manage, and the slightest squeak of the tiniest mouse is enough to set her antennae aquiver. Feline hearing is at its peak in a cat's first three years. After that, it gradually starts to decline.

WHY ARE CATS LIKE ACTORS?

Precious creatures who love to be pampered, cats and actors have a lot in common.
Vivien Leigh adored them. "I've always been mad about cats."
Bette Midler understood them. "Cats always seem so very wise, when staring with their half-closed eyes. Can they be thinking, 'I'll be nice, and maybe she will feed me twice?'"

James Mason admired them. "Cats do not have to be shown how to have a good time, for they are unfailing ingenious in that respect."

Halle Berry thought she was more of a dog person, until she made *Catwoman* and fell in love with the film's cat, Play-Doh, on the set. It wasn't enough to save the film, which one critic described as "gruesome enough to have been ejected from Bill the Cat's esophagus."

Patrick Stewart took home not just $100,000 per episode from *Star Trek: The Next Generation*, but his cat, Bella, whom he found, unwanted, on the set.

⁕ A CAT OF CONVENIENCE ⁕

Visitors to the ladies' restroom in Paddington Station used to be greeted by Tiddles, a vast tabby who lived in the attendant's office. Tiddles never wandered far, and was spoiled rotten by station staff, until his weight ballooned to a disgraceful 29 pounds.

❧ YOUR COUNTRY NEEDS YOUR CAT! ❧

During World War I, the British government was in desperate need of an efficient gas detector to protect troops on the front line from the effects of lethal gas used by the German army.

Five hundred cats were drafted and sent to the trenches, where the unfortunate creatures, like so many of the young soldiers, met with an untimely end. Their hypersensitive sense of smell meant they could detect gas sooner than humans. If a cat suddenly keeled over, it was a warning to the troops to get the gas masks on, as quickly as possible.

In World War II, cats played a safer role. The army food storehouses were a magnet for vermin, and the British Ministry of Supply asked people to volunteer their cats to keep the rat population under control. Cats by the basketful were sent by patriotic pet owners, keen for kitty to do her bit in the war effort.

❧ WHEN CAT EATS CAT ❧

One dark night, in a kibbutz on the edge of the Negev desert in Israel, wildlife guide Arthur du Mosch was awakened by the sound of his pet cat screeching in terror. Leaping out of bed, clad only in underpants and a T-shirt, Mosch must have thought he was still in the midst of some horrible nightmare, when he saw his poor cat dangling from the jaws of a leopard which had crept into the house and was attempting to make off with its dinner.

Du Mosch couldn't leave his cat to such a gruesome fate, and he took a lunge, grabbed the leopard by the scruff and shook it until it dropped his pet. He then manhandled it down while animal handlers were summoned to remove it. Fortunately the leopard was an aged and unusually slow specimen, otherwise du Mosch – and his cat – might not have survived to tell the tale.

Leopards usually avoid humans, although there are tales of them dropping in on unsuspecting families in India. One man had to lock a roaming leopard into his bathroom while he yelled for help, while another household discovered one curled up on their sofa in front of the TV. Conservationist Philip Dowsett has this advice, should you find a leopard in your living room: "Don't run away. Open a door or window as an escape route, and make loud noises to try and scare the leopard away. Whatever you do, don't make a grab. Leopards are extremely powerful, and one well-aimed blow from its paw could kill you."

☙ WHY CATS POOP IN THE WRONG PLACE ☙

Cats are generally clean creatures, but even the most fastidious feline can have the occasional toilet mishap. Isolated incidents aren't much of a problem, but if your cat has begun performing regularly in an antisocial spot, it's time to check out the possible reasons:

- How many cats are there in the household? A dominant cat may take possession of the litter box as part of his or her territory, and chase off the more timid members of the feline community.
- Where do you keep the litter box? Most cats would much rather not take care of their toileting right next to their food and water bowls, thank you very much.
- Has something nasty happened to Montmorency while he was in the box? It's tempting to grab a cat that needs medication, for instance, while he's a sitting target. But if you set up an association of pain or fear with the box, he may never squat there again.
- Is your cat unwell? Urine infections or bowel problems can give Monty a painful stomach, which he'll begin to associate with That Box, and thus conclude that it's safer to go elsewhere. This can be a particular problem for older cats, who are often prone to kidney and digestive problems. Get the vet to give a your pet a checkup.
- Is the litter to his liking? How dare you try to get away with offering cheap, sub-standard litter, which is not to his preferred level of clumpiness?
- Has he grown to love the feel of soft wool carpet under his paws as he communes with nature? If so, you are in deep trouble. Experts suggest this trick: put a carpet remnant in the litter box, and keep Monty away from his favored target areas. Gradually add more, then more litter to the tray, and snip away at the carpet. You need to be dedicated, but if it saves the Aubusson ...

☙ WHY DID THEY ... ☙

... put a cat into the cradle before a baby first slept in it? A Russian superstition, based on the belief that the cat could drive out evil spirits that might harm the newborn.

... keep cats out of the house when someone had died? This was the Chinese, who thought the cats could call back the souls of the dead in the shape of zombies.

... gaze deeply into the cat's eye? The ancient Celts wanted to see the miniature fairy kingdom that dwelled deep within the cat's mysterious eye.

❋❖ JEANETTE WINTERSON ON CATS ❖❋

Author of *Oranges are Not the Only Fruit, Sexing the Cherry, Lighthousekeeping* and many other novels, Jeanette Winterson was born in Manchester, England, and adopted by working-class Pentecostal parents. There were only six books in the house, including the Bible and *Cruden's Complete Concordance to the Old and New Testaments*. Strangely, one of the other books was Malory's *Morte d'Arthur,* and it was this that started Jeanette's life quest of reading and writing.

She works in a "glamorous shed" in her garden, where she can escape from the domestic distractions of "cats, books, pictures, fresh vegetables to cook, the garden to play with, the hens to feed." She once wrote about the cats of Spitalfields:

In Spitalfields, London, where I keep a house, the Romans laid out villas and conduits, and braced the marshy ground for building. They had to be near the river. The river was trade and news and control and escape. The cats that leapt from the boats spread out, keeping near to settlements, as cats prefer to do, but pursuing their own livelihood. The story goes that the force of black cats that used to patrol the Spitalfields fruit and veg market, were the descendants of those Roman cats.

I believe it. When the gigantic wholesale market finally closed and moved out of London altogether, the cats vanished, as though many hundreds of years of work was finally over.

One stayed. Blackie the market cat, moving through the abandoned barrows and haphazard stacks of pallets, belonging to no one, known by everyone, until old age stiffened him.

Friends of mine took him in, and when he died, I buried him in the foundations of my basement. I was digging out the basement at the time, and this act of piety and superstition seemed like a charm. Put the past where it belongs – in the footings of your life – but do it with reverence and memory.

❧ HOW DO CATS FORECAST THE WEATHER? ☙

"THERE WILL BE SCATTERED SHOWERS ..."

In England, cats wash their ears before wet weather, but in
Scotland they rub up against the table legs, in Denmark they tear
up and down stairs, and in China they solemnly wink one eye.

"A FRONT IS APPROACHING ..."

When McWhiskers claws the furniture, say the Scots, it'll soon
be brewing up for high winds. Sailors who toss the ship's cat
overboard get their comeuppance in the storm that will surely
follow. If a ship is stalled in the doldrums, the wind can be raised
again by performing a ceremony with a cat on the ship's bridge.

"EXTREME TEMPERATURES ARE PREDICTED ..."

When puss falls asleep with every limb stretched out, then a
heatwave will soon arrive. But when she's tightly furled in a ball,
with nose tucked under the tail, get your woollies out. A cold
snap is just around the corner.

> *Another cat? Perhaps. For love there is also a season;
> its seeds must be re-sown. But a family cat is not
> replaceable like a worn-out coat or a set of tires. Each
> new kitten becomes its own cat, and none is repeated. I
> am four cats old, measuring out my life in friends that
> have succeeded but not replaced one another.*
> Irving Townsend, American author

⁎⁛ LEONARDO DA VINCI (1452–1519), RENAISSANCE MAN AND ARTIST EXTRAORDINAIRE ⁛⁎

"The smallest feline is a masterpiece."

Leonardo's notebooks are renowned. In them, he wrote copious notes and drew sketches of anything and everything that interested him, writing backwards rather than left to right.

In his notes on *Divisions of the Animal Kingdom,* he categorizes: "The Lion and its kindred, as Panthers. Wildcats, Tigers, Leopards, Wolfs, Lynxes, Spanish cats, common cats and the like."

Leonardo loved animals, always kept pets and became vegetarian in later life, remarking that, "The time will come when men such as I will look upon the murder of animals as they now look on the murder of men."

One of his liveliest sketches was of "27 Cats and a Dragon." The cats are shown in all manner of feline poses – stretching, snoozing, pouncing, either alone, with other cats, or being caressed by human hands. These lively sketches are lovely and reveal the solemn affection he had for felines.

Cats crop up in Leonardo's paintings as well. During the early 1480s, he experimented with a series of drawings of the Madonna, in which the infant Christ is holding a cat, based on a legend about a cat that gave birth to a kitten at the exact moment that Jesus was born. In the studies, the cat is shown in different poses – being cuddled, stroked and even half-strangled.

For a long time experts believed that only these initial studies existed, but then Italian industrialist and art collector Carlo Noya noticed similarities between a painting he owned and Leonardo's sketches. *Madonna with the Cat* was eventually attributed to Leonardo, after being radiographically examined.

What do you call a cat that's swallowed a duck?
A duck-filled fatty puss.

☙ WHY DO CATS LOVE LIBRARIES? ❧

What could be more disarming than to nip into the library and find a large ginger cat zzzz-ing among the scattered newspapers, a black and white cat studying the computer screen intently, or a tabby sitting on the circulation desk, watching the books being checked in and out?

A Web site, dedicated to documenting (and picturing) library cats, notes that worldwide, there are more than 200 cats living in libraries, and bringing pleasure to staff and readers alike.

These figures do have to be interpreted quite liberally. Along with live and late felines, the United States library-cat tally includes: 22 statues of cats, two stuffed lions, one stuffed Siberian tiger, one stuffed cheetah and a ghost.

How do cats come to be living in libraries? The story of Dewey Readmore Books, to give him his full name, is fairly typical, and goes like this. Abandoned kitten turns up on library doorstep – or, in Dewey's case, is dumped into the overnight book-return bin. Soft-hearted staff can't resist, plead with the powers-that-be and adopt. A weekly collection covers the cost of food, a rotation is set up for litter-box duty and soon the kitten is as much part of the scenery as the shelving.

So it was with Dewey. Staff held a contest to choose his name (and a very appropriate selection it was, as the Dewey Decimal System is a method of book cat-egorization), and put Dewey on to the official staff. Among the duties on his job description (yes, they did take it that seriously) were:

- Reducing stress for all humans who pay attention to him.
- Sitting by the front door every morning at 9 a.m. to greet the public as they enter the library.
- Sampling all boxes that enter the library for security problems and comfort level.
- Providing comic relief for staff and visitors whenever possible.

- Climbing in book bags and briefcases while patrons are studying or trying to retrieve needed papers underneath him.

Many people visited the library just to see Dewey, who reigned over the book stacks for 18 years, until his sad demise. In tribute, the library staff said, "Dewey will be remembered by the thousands of people whom he cheered, simply by being a loving presence in the library."

⁙ OTHER LIBRARY CATS ⁙

RODGER THE LODGER
The brindled Roger the Lodger fits in well with the tastefully beige color scheme at the Fort Langley Library in British Columbia, Canada.

RED
Red, a black cat (obviously) of the Ewart Library, Dumfries, Scotland, is a bit of a computer buff, judging by the amount of time he spends gazing at the screen.

CHARLIE
Charlie, who haunts the library of the Whanganui Universal College of Learning in New Zealand, takes a dim view of intellectual endeavor, and prefers to spend his time snoozing on top of the printer.

PRUNELLA
Prunella, a neat white-and-black puss, tiptoes through the square-tiled floor of the Central Architecture Library of Turin, Italy.

LIBRIS
Libris is a very pretty silver tabby who insists that visitors to the Newtown Community Library in Wellington, New Zealand, stop looking at all those tedious books and cuddle her instead.

LOEDER
Loeder is a neat ginger, who just fits into the in-tray of the Amsterdam Library in the Netherlands.

EMILIA

Emilia, black and fluffy with glorious green eyes, prowls the shelves of the Ventspils library in Latvia.

SIMPKINS

Black-and-white Simpkins seems oblivious to the splendor of his surroundings as he stalks the college courtyards around the Hertford College Library in Oxford.

TIGGY

Tiggy has a bed next to the filing cabinet in Holbeach Library, Lincolnshire.

NERO AND HIS SUCCESSORS

The first cat to live in the Veterinarian Library in Zagreb, Croatia, was named Nero, and his three successors have all been named Pero.

ROGERS AND OLLIE

Rogers and Ollie – one ginger, one black and white – keep each company in the British Library in Wetherby, West Yorkshire.

SOLLY

Pure white Solly used to take the scenic route across the computer keyboard in the Gannawarra Library in Australia, rather than go via the floor.

PIKSA

Piksa likes lending a paw with shelving, in the Technical Library in St Petersburg, Russia.

FUNI

Funi is a cute little ginger cat, who knows how to get visitors to the National Library of Iceland in Reykjavik saying, "Aaah." He just crams himself into the branches of a potted plant, and drops off to sleep like a miniature lion.

Cats are cleverer than dogs. You can't get eight cats to pull a sledge through the snow.
Anon.

❧ HOW MANY TYPES OF TABBY? ❧

Just as all cats look alike in the dark, all cats are tabbies in disguise, unless clever breeders have successfully manipulated a breed's coat pattern to eradicate any trace of tabbiness. The tabby coat pattern comes down in the genes from their African wildcat ancestors. If you look closely at a tabby's coat, you can see that in between the characteristic black stripes or spots, are patches of what is known as "agouti" hair. Examine an individual agouti hair – go on, pick one off your dress now – and you'll see that the base is paler than the tip. En masse, the agouti hairs give a mottled, salt-and-pepper effect, very useful in the wild for cats who want to blend in with light and shadow.

You can sometimes detect the faint ghost of tabby markings on a black cat if you inspect it carefully in a good light. It's most obvious in black kittens, and may disappear as they mature.

Not all tabbies look exactly the same. There are four basic ways to be a tabby:

MACKEREL OR STRIPED

One stripe runs along the spine, and slender bands radiate from it, down the sides, rather like the bones on a fish skeleton. This was the most common pattern in Europe until the last couple of hundred years, when it was overtaken by:

CLASSIC OR BLOTCHED

Broad stripes swirl out from blotches on the flanks. This pattern is common in North American and Australian cats, and was probably exported from Britain during the 18th and 19th centuries.

TICKED OR ABYSSINIAN

The head may be the only part that's clearly striped, while the rest of the body shows subtle dark flecks. Contemporary African wildcats often have all-over ticking, and this pattern seems to have spread east into Asia, while the stripier variants went north toward Europe. The pattern is common in Sri Lanka and Malaysia. The ticked pattern is especially striking in the Abyssinian, which resembles the cats in ancient

Egyptian images, and was first brought from Abyssinia (now Ethiopia) to Britain in the late 1860s. The breed has clearly marked "frown" lines on the face, but no stripes elsewhere on its body. The rest of its coat is subtly ticked with darkly banded hairs distributed evenly through a lighter background.

Spotted

The spots are actually broken stripes, and may follow the lines of the mackerel pattern or appear random. Spotted tabbies often have stripy tails and legs.

✺ WHY WON'T MY CAT EAT FOOD FROM THE FRIDGE? ✺

Because freshly-killed meat is warm. This behavior harks back to the wild, where the best way for a cat to tackle a meal is by ripping it apart the moment it's been killed, before some other predator can snaffle it. You can store a half-used can, or homemade food in the fridge, but let it come to room temperature before you offer it to your cat. Throw any uneaten food away, as it will quickly turn rancid if left in the bowl.

✺ WHO GETS FIRST STROKE? ✺

In a two-or-more cat household, the resident felines may compete for your attention. If you detect an element of competition, be careful how you stroke the cats, because you might just make matters worse. If you stroke one cat and then another, you will transfer the first cat's scent on to the second – who won't be made at all happy. To be on the safe side, stroke one cat with each hand, or stroke one cat, then wash your hands before stroking another.

✺ DO I NEED TO CUT MY CAT'S CLAWS? ✺

If your pet spends plenty of time outdoors, she'll wear down her claws naturally through climbing, scratching trees and fences and generally prowling around. The claws of older, less active cats, and those who live indoors will gradually get longer and longer, until you notice your pet getting hooked up on carpet or other fibers.

It's possible to cut your cat's claws yourself, although if she puts up a fight you're better off leaving the job to the experts. You can buy claw clippers from a vet or pet

supplier. Hold the cat on your lap, grasp the paw confidently and squeeze gently so the claws are extended. Then clip, making sure you don't take the claw off too close to the quick; otherwise it could bleed. Take off too little, rather than too much.

HOW CAN I GET MY CAT INTO ITS PET CARRIER?

Be bold, be fast and be wily! Try to shield your pet from any signs that "something's going on." Cats know when a vacation is in the air. They seem to recognize suitcases and have an unerring habit of settling down on a pile of clean laundry that's about to be packed. If they detect any hint of pre-travel anxiety though, they'll be off, so try to keep the preparations as calm as possible.

If there's any risk that your cat might make a break for it, keep the cat flap closed for 24 hours before travel. Confine your cat to a room for an hour or so before you wish to leave for the pet-boarding facility, and let him settle down. Keep the pet carrier out of sight (and out of hearing range, too, if it's a basket that creaks). Get the carrier open and all ready to receive your pet, then quickly pick him up, pop him in and close the lid before he's realized what's going on.

SHOULD MY CAT DRINK MILK OR WATER?

Cats don't need to drink cow's milk, and some are lactose intolerant and may suffer from diarrhea if they do have it. If milk doesn't upset your cat's insides, then you can give the occasional saucer, but don't make it a daily routine.

Cats are quite fussy about what they will drink (no surprise there!). Your Holly is quite likely to turn up her nose at a bowl of fresh, clean tap water, because she can smell the chlorine and other chemicals used to treat it. No thanks! She'd rather have natural, standing water from a muddy puddle or pond. Domestic water that's stood for a while and lost its chlorine smell is more appealing than what comes out of the tap, which is why some cats will risk lapping at the water in the toilet bowl, or will take a drink from the fish tank or flower vase. Discourage this by keeping the lid of the toilet down (yes, men, and the seat!), covering tanks and placing vases out of reach.

That said, some cats seem fascinated by water from a dripping tap, and will enjoy drinking it. They're probably attracted by the light that sparkles from the drips.

●● SHOULD I LET MY CAT HAVE A BARF? ●●

Sounds horrible, but BARF stands for Bones and Raw Food Diet, a new method of pet feeding which has found its way to the United Kingdom from the United States over the past couple of years.

The raw diet developed in response to owners concerned about the high-carbohydrate content often used to bulk out manufactured food, particularly cheaper brands, the amount of additives in the food, and the fact that it is cooked during processing, while cats evolved to eat raw.

A BARF diet tries to replicate the kind of foods cats would consume in the wild – 90 to 95 percent fresh raw meat, vegetables, fruit, nuts and seeds. And no, you don't have to do all the preparation yourself, as you can buy frozen raw foods for your pet.

CATS AND COMPUTERS 4
Why call it a mouse if you can't chase it?

●● COLLARED! ●●

Don't put a collar on your cat just to make him look smart, or to deal with fleas. There are other ways to tackle fleas that are more effective, and collars have dangers in themselves.

The best use of a collar is to equip Buster with some form of ID, just in case he takes a wrong turn and gets lost, is injured and can't get home, or makes a unilateral decision to find a new home.

A capsule with his details on a slip of paper inside is one possibility, but the top can be extremely hard to remove, or else so loose that he comes home without the cylinder. Writing can fade, so use a black pen and print clearly.

A metal disc engraved with address and phone number is more durable, but some cats don't like wearing them, and if the collar gets lost you'll have to get a new disc engraved.

Choose his collar with care. Although collars with elastic inserts mean that if Buster gets caught up he'll probably be able to wriggle free and won't be throttled, these collars do present other dangers, especially when the elastic becomes worn and loose.

- Cats can get their front leg through the collar and get stuck. Sounds unlikely, but actually it's easily done, and if you notice your cat limping on a foreleg, this is the first thing to check. If the trapped leg isn't spotted for a while, the cat can end up with a nasty and painful injury in the "armpit." And worst of all, if the leg gets stuck while Buster's out and about, he may not be able to walk well enough to get home again.
- Elastic collars can also get caught in a cat's mouth, or on his teeth. Nasty.
- Because these collars are designed to slip off easily, your cat will regularly lose both collar and identifier.
- You can buy collars that snap open if they're pulled. These avoid the potential strangulation problem, but again, you can expect to go through quite a few if your cat's fond of exploring in the undergrowth.

Whether or not Buster sports a bell on his collar is up to you. The idea is that it warns unsuspecting birds that someone is creeping up on them. Cunning cats – and what cat isn't crafty? – soon learn to keep their heads still as they sneak towards their prey, to silence that giveaway jingle. But the Royal Society for the Protection of Birds has found that some birds are saved if known hunters wear a bell.

> *"Those who will play with cats must expect to be scratched."*
> Miguel de Cervantes, 16th-century Spanish novelist and playwright

HOW MUCH, HOW OFTEN?

Kittens need lots of tiny meals, to keep their little tummies full. Adult cats are better off with one or two meals a day.

Exactly how much food depends on Simba's age and how active he is. A rule of thumb is around three-quarters to a whole large can of food, or the equivalent, per day.

The danger when feeding several times a day is that you give too much each time. Work out the right amount for the day, then divide into smaller amounts for each meal.

It's worth buying special foods for kittens, who need a food that's higher in protein in fat while they're growing so fast.

Senior cats can also benefit from a food designed for the mature pet. They may have less appetite than before, and need a good, energy-rich diet, which contains easily digested protein and fat.

✺ WHO IS TABITHA LARK? ✺

What a fitting name for a cat. It belongs to a fluffy cat who lives in the Jamieson Library of Women's History in Penzance, Cornwall, England. Tabitha is named after a character in Thomas Hardy's lesser-known novel, *Two on a Tower*.

Librarians at this establishment are fond of Hardy, and other feline residents have been named Emma Hardy after the novelist's first wife, and Florence Hardy, after the second Mrs. H. Other cats – there've been a few – have been named for characters out of Hardy novels, including Bathsheba Everdene (heroine of *Far From the Madding Crowd*), Arabella Donn (first wife of Jude in *Jude the Obscure*) and Grace Melbury, a small fluffy cat who enjoys digging around in the library's potted plants and was christened in honor of a character in Hardy's *The Woodlanders*.

✺ CAN A CAT COME BACK FROM THE DEAD? ✺

- Eighty days after a Taiwan earthquake in 1999, a cat was pulled from the rubble of a collapsed building. It made a complete recovery.
- A microchipped cat, Ted, jumped out of a window and vanished. Ten years later, he was returned to his astonished owner.

What do cats enjoy on a hot day?
Mice cream.

✺ AMAZING FACTS ABOUT THE FELINE ANATOMY ✺

- Jake had a grand total of 28 toes – seven on each paw. He lived in Ontario, Canada.
- A Finnish Maine Coon, Mingo, had whiskers measuring 6.8 inches long.
- Tinker Toy of Illinois, a Himalayan Persian, was the tiniest ever domestic cat at 2.75 inches tall, seven inches from nose to tail. He weighed less than a bag of sugar, at just 1.5 pounds.
- The placid Ragdoll is the largest breed, and can weigh up to 20 pounds. The Maine Coon is also very large.

- You could fit two Singapuras on your lap. Females of this miniature breed can weigh as little as eight pounds.
- A cat can have 100 kittens in its lifetime – a good reason for neutering. Left to their own devices, one pair of cats and their kittens could multiply to 420,000 cats in just seven years.
- A chocolate Burmese, Tarawood Antigone, gave birth to the largest-ever litter of 19 kittens in 1970. Four were stillborn. A South African Persian, Bluebell, gave birth to 14 kittens in one litter, and all survived.
- Dusty had 420 kittens during her life. Litty had 218 kittens, and gave birth for the last time when she was 30.
- A cat's nosepad has a unique ridged pattern, like a human fingerprint.
- Your cat's heart beats twice as fast as yours, at 110 to 140 beats per minute.
- The domestic cat is the only feline that holds its tail vertically as it walks. Wildcats hold their tails horizontally, or keep them tucked down.
- Cat's bodies have 290 bones and 517 muscles. The feline vertebrae has five more bones than a human's. A cat's shoulder blades rise above the spine when it walks, just as they do in a strolling lion or tiger. Flexible carpal bones in the front legs are like human wrists, and give cats the dexterity they need to fish their prey out from an inaccessible nook, or to tiptoe along the fence-top.
- When a cat is resting, its claws are automatically sheathed by muscles in the paw. It can flick them out instantly by flexing the tendons in the leg.
- Within a cat's tail is a fine string of articulated vertebrae.
- What does Archibald have in common with a giraffe and a camel? He walks by moving first his front and rear right leg – then the front and rear left leg.
- Never give your cat dog food. She needs five times more protein than her canine companions.
- It takes time to have a tongue bath. Cats spend about one-third of their waking hours grooming themselves.

❧ WHAT HAPPENED WHEN MOG FORGOT ABOUT THE CAT-FLAP? ❧

Generations of children enjoyed the moment when the lovable tabby star of the *Mog* series of books squashed all the flowers in the window box and made Mrs. Thomas say, "Bother that cat!" Mog is a plump dimwit of a cat, much loved by her owners – and her young readers – even when she's being totally exasperating.

The early life of Mog's creator, Judith Kerr, was incident-packed. Born in Berlin, her father was a drama critic and a distinguished writer whose books were burned by the Nazis, and Judith escaped with her family from Hitler's Germany in 1933 when she was nine years old. The family passed through Switzerland and France before finally arriving in England in 1936. Judith went to 11 different schools, worked in the Red Cross during the war, and won a scholarship to the Central School of Arts and Crafts in 1945.

In the final book in the series, *Goodbye Mog,* the elderly tabby feels so tired she could sleep forever – and so she does. But a little bit of her stays awake to keep a watchful eye on the Thomas family, who miss her terribly. Then, one day, a kitten wanders into the house. A stupid kitten, thinks Mog, and gives the new arrival a helpful paw into Debbie's lap. And so the family have a new pet to love, but: "We'll never forget Mog," say the children. "I should think not," says Mog, and she flies up and up, into the sun.

What's a cat's favorite film?
The Sound of Mewsic.

❧ ARE HUMANS MAKING CATS ILL? ❧

Your pet can suffer from the effects of secondhand smoke just as much as anyone else in the household. Cats who live with smokers are more likely to develop certain cancers. And if your cat starts to cough and splutter, he could be suffering an asthma attack brought on by dust, smoke or even human dandruff.

❧ CAN CATS DEVELOP ALLERGIES? ❧

Do your friends start sneezing as soon as puss strolls into the room? It's not unusual for people to be allergic to cats, but did you know that cats can develop allergies, too?

INSECT ALLERGIES

Many cats are sensitive to fleas, and come out in a rash at the bite-site, or even lose small clumps of fur. Signs of a flea infestation are frantic scratching and flecks of black dirt that you can see among light-colored fur, and which is dropped wherever your cat sleeps. The first step is to ask your vet for an allergy treatment to heal any broken skin and calm the itching. You must also treat your cat for fleas. Prescribed remedies are more effective than powders bought over the counter. Try one of the liquid medications, which is applied to the cat's nape. You should also vacuum your home thoroughly and spray with anti-flea spray, to make sure your cat can't be re-infected.

ALLERGIES TO AIRBORNE SUBSTANCES

Yes, cats can get hay fever, and the symptoms are just like yours – runny eyes and sneezing. Vets can prescribe medication, and it's probably kindest to keep your cat inside when the pollen count is high. Along with pollen, cats can be sensitive to spray cleaners used around the house, so if you notice your cat rubbing its eyes or sniffling after you've been cleaning, switch to other products.

CONTACT ALLERGIES

Common houseplants such as the rubber plant can trigger an allergic reaction in your cat. They can also respond to house dust, wool (including carpets), newsprint, household products and even some types of cat litter. Symptoms to watch out for are spots and sore patches, which are easiest to see on the chin, ears, tummy, inner thighs and under the tail. Your vet can prescribe medications to help, and cats can even have skin patch tests to track down the cause of their discomfort.

FOOD ALLERGIES

Think twice before you offer your pet a morsel of your supper. Vets advise against it, and certainly chocolate and some dairy products can cause tummy upsets. Cats who become allergic to turkey or chicken will reject pet foods that contain poultry. If your cat develops a serious food allergy, your vet may be able to prescribe special foods that are safe to eat.

Cats and monkeys; monkeys and cats; all human life is there.
Henry James, American novelist, author of *The Turn of the Screw*

❧ WHO ARE THE LONGHAIRED LOVELIES? ❧

Adorably fluffy, the longhaired breeds, known as Persians in North America, were first known in Asia, and were probably brought to Europe in the mid-16th century. The Victorians adored them, and in the 20th century, dozens of different longhaired breeds were developed.

Now You See It ...

Look at the Balinese in winter, and you see a longhaired version of the Siamese. The distinctive slender, long-legged cat wears an exquisite silky coat and has a gracefully plumed tail. In spring quantities of fur are shed, so the cat looks much more like a shorthair. These cats are easy to groom, because the flat, silken fur doesn't mat like the coats of fluffier breeds.

COLD-WEATHER CATS

The Maine Coon is a big, shaggy cat that can weigh as much as 22 pounds. Its luxurious coat flows over its solid, well-muscled body, and its tail is marvelously bushy. In 1861, the first Maine to be shown in North America was a black-and-white named Captain Jenks of the Horse Marines.

You can just imagine the Norwegian Forest Cat at home in the depths of a frozen Scandinavian fastness. Its ancestor, the skogkatt, or forest cat, really did live in the forests of the Norselands. It wears warm and downy under-fur, covered by a spun-silk, non-tangling outer coat that repels the weather, and is shed in the warmer months.

FURRY FISHERMEN

The Turkish Van is a rare and ancient breed that originates from the Lake Van region of Turkey. It was brought to the United Kingdom in 1955, although it didn't head to North America until 1982. This cat stands out from the crowd because of its gorgeously silky longhaired coat, which is predominantly white, with colored markings confined to the bushy tail and head. The fur is water-resistant, and perhaps because of its lakeside background, the Turkish Van is called the Swimming Cat in its homeland, and enjoys a cooling dip.

COULD A CAT CALL THE POLICE?

Maybe. Police rescued a man who'd fallen out of his wheelchair and couldn't reach his alarm buzzer, when they were summoned by an emergency call. Question was – who had dialed? The man was stretched out immobile on the floor, and the only other person at home was his ginger cat, Tommy, who was found crouched by the phone.

❀❖ MICROCHIPPING ❖❀

A tiny chip, about the size of a rice grain, is inserted painlessly under the skin of the nape. You register a code for your cat, which is recorded on the chip and can be read with a hand-held reader. Simple.

But everything depends on the person who finds an injured, or, don't-think-about-it, dead cat taking the initiative to call a rescue center and get the creature checked for a chip.

Too often, people assume that a cat with no visible means of ID isn't owned and cared for, and while they would look at a collar and make an effort to find the owner, they won't investigate to see if the cat has been chipped.

Is it bad luck if a black cat follows you?
Depends if you're a man or a mouse.

❀❖ THE TRUE STORY OF A VERY FAT CAT ❖❀

Hercules left home while his owner was seriously ill in the hospital and disappeared off the face of the Earth.

The cat-sitter was distraught and scoured the neighborhood, but with no luck. She assumed that Hercules must have come to a bad end. But as it emerged, his end was not so much sticky, as stuck.

The crafty cat had set up home in a nearby garage, and been feasting off the family's store of dog food and getting fatter, and fatter, and fatter. The tubby tabby had ballooned so much that he eventually came to light when the astonished owners found him jammed firmly in the doggie-door, victim of his own over-sized appetite.

Hercules was extricated from the dog-flap and taken to the local pet refuge, where staff nicknamed him Goliath because of his gigantic girth. The story has a happy ending, because when the local TV station ran a feature on Goliath, his owner, now home from the hospital, saw the piece and reclaimed his pet, still recognizable, but somewhat larger than when they'd last met.

It's better to feed one cat than many mice.
Norwegian proverb

❀❁ ANNE FRANK'S CATS ❁❀

In Anne Frank's astonishing diary, kept from the time she and her family entered the sealed-off back rooms of her father's Amsterdam office building in 1942, until their betrayal to the Nazi Gestapo in 1944, she simply but movingly records everyday life in hiding. In the midst of a life of fear, were little oases of pleasure, some of them provided by cats.

July 8, 1942, the day before going into hiding
Anne's family make their secret preparations to leave their home. The only person to whom Anne, then aged 13, says farewell is Moortie, her little cat, who went to a good home with the neighbors.

August 14, 1942
Peter van Daan, 15, arrives in the Secret Annex, bringing his cat, Mouschi, with him.

November 17, 1942
Anne writes a tongue-in-cheek "Prospectus and Guide to the Secret Annex," stating that small pets are permitted and that good treatment is available.

August 18, 1943
When Anne goes up to the attic to fetch some potatoes, she finds Peter de-lousing the cat. As he looks up, the cat darts out through the open window and sits out of reach in the gutter, which amuses Anne but perplexes Peter.

August 20, 1943
The cat, of course, is oblivious to the danger of being seen. One night, it's Peter's job to creep into the main office to collect the bread. He crawls past the windows so as not to be seen from outside when suddenly Mouschi bounds playfully over him and sits under a desk. Peter has no choice but to retrieve the cat, and has to crawl back into the office and drag the animal out by the tail, having failed to tempt him with a piece of bread.

January 24, 1944
Moffi, the warehouse cat, has been growing plumper, and Anne wonders if there are kittens in the offing. Peter points out that he's a tom, and turns the cat over and displays his underside to Anne, to demonstrate just how masculine Moffi is.

May 3, 1944

A sad day in the Secret Annex. Moffie has gone missing. Anne's streak of realism comes to the fore. These are desperate times, and she reflects that Moffie may have become someone's meal or his skin made into a fur cap.

May 10, 1944

The sound of water pattering, and upstairs Mouschi had shunned the litter box and peed on the floor. The drips through the ceiling form a pool in and beside the precious barrel of potatoes. Chaos ensues as Mouschi, sensing he is in the wrong, crouches under a chair, while Peter wields water, bleaching powder and a cloth. Anne thinks it all very funny.

May 19, 1944

By now Anne is 15, and she and Peter are growing much fonder of each other. She tells how Peter blushes when she kisses him goodnight and wonders if she is a good substitute for Moffi.

Anne's last diary entry was made on August 1, 1944. On August 4, the SS raided the Secret Annex, and Anne was eventually sent to Bergen-Belsen concentration camp, where she died of typhus on April 12, 1945.

What do cats like for breakfast?
Mice Krispies.

❧ AN APOCRYPHAL TALE? ❧

What we can only hope is an Internet legend concerns the largest cat, Verismo Leonetti Reserve Red, known as Leo, a Maine Coon weighing 35 pounds and four feet long from nose to tail and the world's smallest cat, two-pound Mr. Peebles, a blue-point male Himalayan.

The pair both held Guinness Book World Records for longest and smallest cat, respectively, and met for a publicity shoot. Whereupon, when the humans' backs were turned, Leo ate Mr. Peebles, munch, crunch.

❦ WHY SO MANY COAT COLORS? ❦

The ancestor of all domestic felines, the African wild cat, was shorthaired with tabby markings, perfectly patterned for disguise in its natural environment. Most coat patterns still reflect these origins to some extent, although in a few breeds that distinctive tabbiness has been completely bred out.

As cats spread around the world, the first major change in their coloring was probably the evolution of the solid black coat, often found in jungle-dwelling cats. Other genetic mutations created ginger and white cats, and these types survived because, once domesticated, cats didn't need to be so well camouflaged as they were in the wild.

It's from this small basic range of shades that the massive variety of feline fur colors originates. In any neighborhood, you'll be likely to find cats with fur patched in black-and-white, white-and-black, or ginger-and-white. Pure tortoiseshells – a mix of black and red, sometimes with the addition of white – are less common, and the complexities of feline genetics mean that most are female. If these cats have traces of tabby markings as well, they're known as tortie-tabbies, or torbies.

More exclusive are the instantly recognizable pointed patternings, best known on the Siamese, where the lighter shaded body coat is set off to perfection by a darker color on the tail, legs, paws, face and ears.

⋄ WHAT'S GOING ON 'EAR? ⋄

A quick look at your cat's ears will tell you what kind of day she's having.

- Straight up? She's on the alert, ready for anything.
- Pointed forwards? Something's piqued her feline interest and she's just figuring out exactly what it is.
- Flat back to the skull? Help! She's had a nasty fright.

⋄ CATS WITH RINGLETS? ⋄

Yes, they do exist, in the Rexed breeds of curly-coated cats.

SELKIRK REX

The Selkirk Rex is the curliest of all. It looks as if it's wearing its very own astrakhan coat, because its body is covered with artfully disheveled curls. The very first specimen of this breed was a curly-coated rescued cat taken in by a breeder in Montana, who named her Miss De Pesto – Pest for short – and mated her with a champion black longhair named Photo Finish of Deekay. Three of their kittens had curly coats, and the breed was developed from them.

Litters from these cats can include a mix of straight and curly coats, and breeders can tell which are which right from birth because the kittens who will grow up curly-coated have corkscrew whiskers.

LA PERM

The aptly named La Perm can have anything from a lightly waved, rather fluffy coat, to a full set of tight ringlet curls. The curliest fur is on the belly, and around the throat and ears. This breed began in 1982, when a farm owner in Oregon noticed a bald kitten in among her cat's litter. After a few weeks, this kitty began to sprout a soft, curly coat, and was immediately named Curly. Over the next five years, several more curly-coated kittens were born on this farm, and eventually the owner consulted the cat-breeding experts, and learned how to control the breeding and produce a new pedigree.

CORNISH REX

West Country wonders, the Cornish Rex and Devon Rex are close neighbors who look similar, but actually spring from two different genetic mutations. The Cornish Rex has a rippled coat that looks like corrugated cardboard, but feels like cut velvet.

The first of this breed, Kallibunker, was born to Serena, a farm cat who lived near Bodmin, Cornwall. The cat has a distinctly Oriental look, with large ears, a slender, tapering tail and long, elegant legs. It's a very athletic breed, and thinks of nothing of bounding from the floor to its owner's shoulder to say hello!

DEVON REX

Looking like a startled pixie, with its elfin face, pointy ears and huge eyes, the Devon Rex has a very short, soft coat that lies in stroke-able ripples or swirls. Breeders have nicknamed this breed "the poodle cat," and describe the personality of this curious and energetic creature as being a like a cross between a cat, a monkey – and Dennis the Menace! The first specimen, Kirlee, was probably fathered by a curly-coated stray that Devon locals had noticed hanging around.

I am fond of pigs. Dogs look up to us.
Cats look down on us. Pigs treat us as equals.
Winston Churchill, British
wartime Prime Minister

❀ WHICH CAT WON A SILVER MEDAL FOR BRAVERY? ❀

Faith, a little tabby stray who made herself a home in St. Augustine's and St. Faith's Church in London in 1936.

Four years later, in August 1940, Faith gave birth to Panda, a black-and-white kitten. On September 6, she began to behave strangely, picking Panda up by the scruff, and carrying him down into the basement, a dank and chilly area, littered with assorted junk and piles of old sheet music. Three times, Father Ross, the rector, brought the pair of them back upstairs. Three times, Faith insisted on carting her tiny son back down to the basement.

On September 7 and again on the 9th there were two huge air raids with hundreds of casualties. On the night of the 9th a bomb fell on the church, and in the morning all that was left was a smoldering ruin with a mangled roof that was threatening to cave in.

Although firemen warned him to stay clear, Father Ross couldn't bear to abandon Faith and Panda without trying to find them, and he clambered over the wreckage, calling them. He heard a faint meow and began to dig through the ruins with his hands, throwing piles of burned sheet music to one side. There, squeezed between two piles of papers, was Faith and, huddled underneath her belly, Panda.

Father Ross took the two cats to safety, and shortly afterwards the roof collapsed on the spot where they'd been sheltering.

He was so staggered by the cats' survival that he took a photo of Faith and pinned it up, with this notice:

"Faith"
Our dear little church cat of St. Augustine and St. Faith.
The bravest cat in the world.
On Monday, September 9th, 1940,
she endured horrors and perils
beyond the power of words to tell.
Shielding her kitten in a sort of recess in the house
(a spot she selected three days before the tragedy occurred),
she sat the whole frightful night of bombing
and fire, guarding her little kitten.
The roofs and masonry exploded.
The whole house blazed. Four
floors fell through in front of her.

Fire and water and ruin
all round her.
Yet she stayed calm and steadfast
and waited for help.
We rescued her in the early morning while
the place was still burning, and
By the mercy of Almighty God, she and
her kitten were not only saved, but unhurt.
God be praised and thanked for His goodness
and mercy to our dear little pet.

As more people saw the notice, Faith's story began to spread, but it wasn't until 1945 that Rosalind, Father Ross's wife, read a newspaper article about a dog that had been awarded the Dickin medal, and got in touch with Maria Dickin.

Faith couldn't be awarded the medal because she was a "civilian" cat. But Mrs. Dickin was so impressed by the cat's bravery that she had a special silver medal struck for her, and presented the medal to Faith herself, in the presence of the Archbishop of Canterbury. The citation read: "For steadfast courage in the Battle of London, September 9th, 1940."

Panda eventually went to live as the resident pet at a nursing home, and Faith stayed at the church, and lay at her master's feet as he preached his sermons. She died peacefully in 1948.

HARRIET BEECHER STOWE (1811–1896) ON CATS

Her famous book *Uncle Tom's Cabin* contributed to the outbreak of the American Civil War by bringing the evils of slavery to the attention of Americans more vividly than any other book before.

Her cat, Calvin (also her husband's name), arrived on her doorstep one day. She took him in, and he quickly became central to the household, and would sit on the author's shoulder radiating calm, "during hours of frenzied writing." He could open doors, enjoyed his food and took dinner with the family in the dining room. When Harriet went away, a friend who looked after Calvin was impressed by the Maltese cat's intelligence, and said: "He is a reasonable cat, and understands pretty much everything except binomial theorem."

❧❧ HOW CAN I MAKE MY HOME SAFER? ❧❧

Put away small objects. Cats are like little children when it comes to having an unhealthy curiosity over little items, except instead of jamming them up their nose, they tend to eat them.

Objects of interest to cats, which if eaten could at best give a nasty stomach ache, and at worst prove fatal, include:

• elastic bands, hair ties, etc.

• needles and pins

• safety pins

• dental floss

• small coins

• bits of string, ribbon or knitting yarn

• earrings

• tiny plastic toys

A cat could strangle itself in trailing cords from blinds, curtains and lamps, so bundle or tie them together and keep out of reach. Electrical cords are lethal if your cat bites through them, or he might bring down a heavy or hot object, such as an iron, on top of himself. Unplug items and put the cord out of sight.

✿❖ WHAT'S THE PERFECT CAT FOOD? ❖✿

Turn away now if you're squeamish. A really great source of feline nourishment is mice. Nature planned it that way. A meal-in-a-mouse contains protein, fat, vitamins, minerals, some useful furry roughage and few delightfully crunchy bones for healthy teeth. If your cat regularly catches and consumes mice or other small rodents, he needs less food than the somnolent house cat.

How can you tell if your cat's caught a cold?
He has cat-arrh.

✿❖ WHAT IS A PATSY? ❖✿

The Patsy – Picture Animal Top Star of the Year (cinema) or Performing Animal Television Star of the Year – Award was set up by the Hollywood office of the American Humane Association in 1939. The awards are given in four categories: canine, equine, wild and special. And guess who's in the "special" section? Alongside goats and pigs are feline film stars including:

- Rhubarb, who starred with Audrey Hepburn in *Breakfast at Tiffany's,* playing the no-named Cat.
- Clarence the cross-eyed lion, who was beaten into third place one year by a pig and bear, for his role in the TV series *Daktari.*
- Syn Cat, Siamese star, alongside Hayley Mills, of the 1965 Disney classic *That Darned Cat.*

✿❖ JUST HOW MUCH DO CATS UNDERSTAND? ❖✿

More than you might give them credit for, thought 17th-century poet Samuel Butler, who made this observation:

"If you say 'Hallelujah' to a cat, it will excite no fixed set of fibers in connection with any other set and the cat will exhibit none of the phenomena of consciousness. But if you say 'M-e-e-at,' the cat will be there in a moment, for the due connection between the sets of fibers has been established."

CAT/HUMAN AGE CONVERSION CHART

Cat years	Human years	Cat years	Human years	Cat years	Human years
1 month	5–6 months	3 years	28 years	12 years	64 years
2 months	9–10 months	4 years	32 years	13 years	68 years
3 months	2–3 years	5 years	36 years	14 years	72 years
4 months	5–6 years	6 years	40 years	15 years	76 years
5 months	8–9 years	7 years	44 years	16 years	80 years
6 months	10 years	8 years	48 years	17 years	84 years
8 months	13 years	9 years	52 years	18 years	88 years
1 year	15 years	10 years	56 years	19 years	92 years
2 years	24 years	11 years	60 years	20 years	96 years
				21 years	100 years

❧ HOW OLD IS MY CAT IN HUMAN YEARS? ❧

Cats grow up fast in human terms, and race from babyhood right through to rampant adolescence, over their first 12 months.

During year two they add another nine human years, and reach their third birthday as a fully-rounded adult – you hope. After that, they develop at a pace equivalent to about four human years to every cat year.

These conversions are based on averages. Some pedigrees, such as the Maine Coon, mature at a much more leisurely rate. Diet and exercise also have an impact, and a fat cat will slow down and show her age sooner than her more slender sisters.

BABYHOOD

- Ah, how kittenhood flies past. By two months, Jethro's already reached human nine to 10 months and is waving goodbye to being a baby.

CHILDHOOD AND TEENS

- For the next three months, he's adding on three human years for every month of his life. By the time he hits his first half-year, he's been through playgroup, nursery and primary school and is at the same stage as a lively human 10-year-old child.
- Puberty strikes at around eight months, when he's like a human 13-year-old – but not for long. By his first birthday, he's in the full flood of adolescence at 15 human years. Watch out for loud music and slammed doors.

Young Adulthood

- The carefree teenage years are soon over, and by the time he's two, Jethro is like a 24-year old man, strong, energetic and deeply interested in the opposite sex. After that, he adds another four human years for every one of his. His 30s are over by the time he reaches the human big 4–0 at cat age six.

Mid Adulthood

- By the time he's eight, Jethro may have lost some of his youthful exuberance, and settled down to a comfortable routine of eating, sleeping and exploring. He's less likely to play spontaneously, but will still chase, bound and leap with just a little encouragement. And when he thinks you're not looking, he'll pounce and pirouette like a much younger chap.
- The human 40s roll by, and by nine he's reached the human equivalent of 52, hitting 60 when he's 11.

Older Adulthood

- A truly mature cat by now, Jethro is aging gracefully, and sleep is high on the agenda. He reaches retirement age, 65, just after he's 12, enters his human 70s in the run up to his 14th birthday, and by the time he hits 16, is the human equivalent of 80 years old.

Old Age

- Cats used to have a life expectancy of 12 to 15, and cats that roam outdoors and are exposed to the risks of traffic, disease and accidents don't always make it beyond this.
- Non-pedigree cats usually have longer lifespan than purebreds. Neutered cats also tend to live longer, perhaps because they avoid the risks of kittening, and the jungle warfare of rival toms.
- More and more cats do live on, happy and healthy, if a little slower than they were. Age spans of 20-plus are not unusual, and the oldest cats on record survived well into their 30s.
- Between 18 and 19, they're like a 90-year-old human. Jethro can expect the president's telegram when he celebrates his human century at 21.

✺ IS IT POSSIBLE TO SPOIL A CAT? ✺

Never! Indeed, for those with an indulgent nature and bottomless wallet, it's possible to treat a much-pampered pet to:
- a fur-trimmed cat basket
- cat massage, or shampoo and blow-dry
- personalized hooded bathrobe
- pet cologne and scented shampoo
- a custom-built cat's house
- jewelled collar
- heated bed
- floor-to-ceiling play and scratch-post center
- cat-sized Santa hat or sailor costume

✺ WHY DOES CATNIP MAKE MY CAT CRAZY? ✺

If you have a patch of *Nepeta cataria* or *grandiflora* – otherwise knows as catmint or catnip – in your garden, you can expect it to be regularly flattened by any passing feline who's sensitive to the plants. *Nepeta cataria* grows to a height of 20 to 40 inches and has white flowers. The *grandiflora* variety is larger and lusher, with blue flowers in summer. Both herbs have quite coarse dark green leaves and are good for filling borders – but only while they're allowed to stand upright.

Some cats are unmoved by a waft from these aromatic plants, but if your Rooney's one of the two-thirds who are turned on by it, it's a very different story. One whiff, and he'll go into a frenzy of rolling, pawing, leg-waving and leaping, rubbing himself fervidly on the leaves or biting them to release the scent. Some cats growl or mew, while others fill the air with rumbustious purring. Remember the restaurant scene in *When Harry Met Sally* ...? It's like watching your cat act it out. Eventually, as the ecstasy passes, they'll flop down on the plant – good thing it's reasonably hardy – and lie there in a state of bliss.

The catnip trip doesn't last long and eventually Rooney will come back from heaven, shake himself and stroll off, although if he passes the same way an hour or two later, he's quite likely to go through the whole rigmarole all over again.

About two-thirds of cats react to catnip, which suggests that the sensitivity passes down through the genes. Cats of both genders respond, whether neutered or not, so it's probably not the hormones leaping into action, even if that's what it looks like.

Kittens and older cats are less interested in catnip, but lions and tigers love the stuff. There are no known side effects, so the only reason to keep Rooney off the flowerbed, apart from saving your plants from death-by-cat, would be if he is very obese or has a known heart problem, when it could harm him to get hyper-excited.

If you don't grow *Nepeta,* your cat needn't be deprived. Buy him a toy scented with catnip. Although he won't go into quite the same spasm as he would with a clump of the real thing, if he's sensitive to it he still won't be able to resist tossing the toy around and chomping on it.

> *A kitten is more amusing than half the people one is obliged to live with.*
>
> Lady Sydney Morgan,
> early 19th-century Irish novelist
> and one of the most hotly discussed
> literary figures of her generation

☙ ARE WE NEARLY THERE? ❧

How do they do it? Some cats appear to have a built-in compass, or maybe they're attuned to the Earth's magnetic field in some way that gives them an unerring sense of direction. No one knows, but there are plenty of tales of felines who've covered hundreds of miles to get to where they long to be.

SUGAR

Sugar couldn't bear to be left behind when her owners moved 1,500 miles across three states. It took her a year to catch up, but she made it.

GRIBOUILLE

Gribouille didn't want to make the shift with his family from central France to southwest Germany. He ran away, and trekked for 21 months, arriving back at his former home half-starved and footsore.

TIBS

Tibs lived at Preston station in the north of England, and one day hopped on the London express. Not surprisingly, she was spotted and put on the first train home, carefully labeled and in the care of the guard.

FELIX

Felix, who crept into an aircraft cargo hold in Germany, spent four weeks there and flew the equivalent number of miles to circle the Earth seven times, taking in London, New York, Rome, Los Angeles, India and Saudi Arabia.

☙ HOW DO I TACKLE A PROBLEM WITH TOILETING? ❧

Finding – or scenting – a pile of cat poop on the carpet is not good for human/feline relations. Cats are very clean creatures, and seldom have a mishap. But sometimes, if they're upset, or as they get old, little accidents do happen. And when there's been one accident, more seem to follow.

Soiling can be a sign of illness, so if you're at all concerned, take Henry off to the vet.

Try covering the litter box. Some cats feel very insecure as they squat, and prefer to be able to perform under the cover of darkness. Some litter boxes come equipped with a plastic dome, or you can improvise with an inverted cardboard box. Just cut a cat-sized door in it.

Fastidious cats turn their noses up at a dirty litter box, but don't clean it too diligently. Henry prefers it to be slightly used because then it smells delightfully of HENRY. Daily cleaning and refilling should be ample for a one-cat household.

Experiment with different cat litters to see if your cat has a preference. If you want to encourage him to use the garden as a toilet, mix some soil into the litter to give him the right idea, then mix some used litter with soil in a suitable patch in the garden, and show him where his new latrine is sited.

Cats often get the urge after they've eaten, so it's sensible to put Henry outside after his breakfast.

Keep feeding bowls well away from the litter box. Cats object to using a box that's near their food – wouldn't you? – and this can drive them to unload elsewhere. You can also place small amounts of dried food in places that have been pooped upon, to put Henry off.

Tempting as it might be, don't yell at your cat or rub his nose in any wrong-doings. He won't understand, and nerves might lead him back to the same spot again. If you catch him red-handed, pop him quickly into the litter box.

❧ WHAT DID JOHN LENNON'S MOTHER HAVE IN COMMON WITH CHARLES DICKENS? ❧

Both thought they owned a boy cat, until …

Elvis belonged to John Lennon's mother, Julia. She was a big fan of Presley, so Elvis he, or rather she, remained, even after giving birth to a litter of kittens in the bottom of the kitchen cupboard.

Charles Dickens (1812–1870), author of numerous much-loved classics, including *Great Expectations, David Copperfield* and *Bleak House,* owned a white cat, William, hastily renamed Williamina when she produced a litter of kittens. Dickens tried gently to dissuade her from parking her family close to his desk, by carrying them back to the kitchen. Not to be deterred, Williamina transported the kittens one at a time back to a cozy spot by Dickens' desk, and they grew up alongside him as he worked. Dickens kept one small female into adulthood. The Master's Cat, as she became known, learned how to put out the Master's reading candle with a swipe of her paw, when she wanted his attention.

❀❖ FISH WITH CARROT, ANYONE? ❖❀

- Japan – as befits a fish-eating nation, cats can enjoy clam- or squid-flavored food. Also on the shelf, fish and egg, fish with carrot and baby sweetcorn.
- Mexico – cats raised on chili con carne will happily lap up spicy foods that would set their European counterparts sneezing.
- Spain – owners adding a little chopped red pepper to garnish their paella, will sprinkle what's left in the cat's bowl.
- Australia – canned food delicacies include lamb with egg and salmon with cheese.
- United States – gourmet cats can feast off canned treats such as White Meat Chicken Florentine in a Delicate Sauce with Garden Greens, or White Meat Chicken with Whipped Egg Souffle.
- Britain – canned cat food has a fantastic range of flavors, alongside the humble rabbit or sardine, including pork, venison, pheasant, quail, shrimp, lobster, prawn and crab. You can also find cat food that contains rice, cheese, milk or eggs.
- Feral cats anywhere will eat what's going, and those that scavenge around the back of hotels and restaurants often have a diet that's dictated by the tourist trade – cake, fruit, breakfast foods such as scrambled eggs, bacon fat, sliced cheese, or noodles, rice and cream sauce.

❀❖ A CAT MAY LOOK AT A KING … ❖❀

… is a cheeky way of saying "I'm just as good as you are."

A political pamphlet with this title was published in 1652, at the end of the Civil War, when Oliver Cromwell was about to become Lord Protector, having defeated supporters of Charles II, son of the unfortunate Charles I, who was executed in 1649.

•❀ WHAT ROLE HAVE CATS PLAYED IN WARS? ❀•

- In the Crimean War, Russian soldiers took their lucky kitten mascots to the battlefield, hidden under their greatcoats.

- In World War II, it was illegal to give valuable milk rations to cats, who had to drink water instead, unless they were sick or on vermin duty in a warehouse, when they could have dried milk.

- Andrew, a kingsize tabby with a white bib and inverted "V for victory" on his nose, was mascot of the Allied Forces Mascot Club, an organization which lobbied for recognition for all animals that served with the allied forces and civil defense services.

- Despite being an old gentleman of 19, Jim woke his owners when their house caught fire in a raid in 1942, and was awarded the Blue Cross Medal, given to animals who had helped to save people's lives.

- Salty and her kitten unwittingly went along on a rescue mission, when they stowed away on an amphibian rescue plane that was going out to save a pilot who had ditched into the sea off San Diego, Calif.

- A cat found on board a German bomber shot down in Wales in World War II became the first feline prisoner of war, and was incarcerated in an animal shelter and named Tiger.

- World War I Admiralty documents record money spent on feeding the numerous cats employed to keep the rats down on board ships.

- The German battleship *Bismarck* was sunk in 1941, and one of the few survivors was the ship's cat, Oscar. He was seen swimming, and was rescued by the crew of the British destroyer, *HMS Cossack*. Five months later, that ship, too, went down – but once again Oscar escaped drowning, and ended up as ship's cat on board the *HMS Ark Royal*. But after little more than a couple of weeks she, too, was torpedoed. No one really wanted Oscar, now known as "Unsinkable Sam," on board their ship, in case history repeated itself yet again, and he spent the rest of his life safely on dry land.

- Convoy the cat lived on board the *HMS Hermione*, where he had a full naval uniform, his own hammock, and was listed in the ship's book. Unfortunately, he went down with his ship when it was torpedoed in 1944.

Why did the cat oil the mouse?
Because it was squeaky.

TOP 10 CAT BREEDS

Which are the best-loved pedigree breeds? This list comes from the Cat Fanciers' Association, one of the biggest pedigree registration organizations in the world.

1. PERSIAN

Top favorite is this gorgeous fluff-ball of a breed. The Persian has a sweet and gentle temperament to go with its soft coat and open, pansy-like face. It's one of the oldest breeds of domestic cat and has always been a very popular show cat.

The Persian has a rounded shape, with stocky legs, a short neck and broad head with small, tufty ears. Its face is flat and its eyes are large and expressive. These cats are playful, but they're not fond of leaping and bounding, and tend to stay away from high places. Persians are bred in an enormous range of colors and coat patterns, from pure white to solid black, through silver, tabby and tortoiseshell. Be warned, though, that if you fall in love with a Persian you're in for hours of patient grooming every day. This is a luxury, but high-maintenance, pet.

2. MAINE COON

In second place is the Maine Coon, a big, rugged American breed with a powerful body, long, silky coat and magnificent tail.

The Maine Coon has had lots of other names, including the American Forest Cat, the American Snughead and the Maine Trick Cat. There are legends about its parentage, and it's even been suggested that the breed is the result of a pairing between a house cat and a raccoon, although in fact this hybrid would be zoologically impossible. Whatever their origins, Maine Coons have always been prized, and were exhibited informally at New England cat shows held in the early 1860s, where they competed for the title of Maine State Champion Coon Cat.

The breed fell out of favor in the early 20th century, eclipsed by the starry good looks of the more exotic Persians and Siamese. A revival of interest started gradually in the 1970s and now the breed is well-known and loved around the world. Traditionally a tabby, the Maine Coon is also bred in other colors. It's a good-tempered, intelligent hulk of a cat which loves to play.

3. THE EXOTIC

The easy-care Persian, this chunky cat has the same stocky body and flat face as its longhaired counterpart, but its short coat is like the softest, most luxuriant velvet, and cries out for you to sink your fingers into it. This coat needs hardly any grooming by the cat's owner because the plush fur doesn't mat or tangle easily.

The Exotics come in a range of colors, including an exquisite smoky gray. They're quiet cats, who seldom "say" much, but have a docile and serene personality and enjoy human company. Exotics like to play as much as any other breed, but they're equally happy sitting in silent feline contemplation of the world.

4. The Siamese

Slender, elongated and gracefully angular, the distinctive looks of the Siamese set it apart. The Siamese's body is light colored, while its extremities are dark. This is known as a "colorpoint" or "points" pattern, and occurs because the Siamese carries a gene which prevents the pigmentation of its coat when its body temperature rises above a certain point. Where the cat is coolest – on paws, tail and muzzle – the dark color develops. On the animal's trunk, where its body is warmer, the fur stays a paler color because the darker pigment cannot develop.

Siamese cats "talk" like no other breed, and have striking – some would say, strident – voices, which they raise at every opportunity. They're super-sociable cats who love to be in on any human action, and have great personalities.

5. THE RAGDOLL

Calmer than calm, this is the cat for you if you want a pet that's laid-back and mild. Developed in California in the 1960s by breeder Ann Baker, in its early days a myth arose that the breed was not sensitive to pain, although this is untrue.

The Ragdoll has saucer-like blue eyes and a semi-long coat with a pointed pattern. When picked up, it relaxes like a floppy toy, although British varieties of the breed don't display this characteristic as strongly as American Ragdolls. The breed has a very gentle nature, and makes a good indoor pet.

6. ABYSSINIAN

A beautiful, sleek creature with a reddish brown, "ticked" coat, the Abyssinian is said to be very good at training its owners. Extrovert, intelligent, fearless and lithe, an Abyssinian is only happy if it has plenty of freedom. Its motto is "Don't hem me in."

Looking as if it's just stepped off an ancient Egyptian mural, one story about its origin claims that the first Abyssinian, Zula, was brought to England by the wife of Captain Barrett-Lennard when she returned from Ethiopia (formerly Abyssinia) in 1868, and was probably given to her by one of the soldiers involved in the Abyssinian conflict that had just ended. A more prosaic version has it that the cat is "more at home on the Thames than the Nile," having been created through selective breeding in England in the late 19th century. Whatever the truth, the Abyssinian is still a cat to cherish, with its suave good looks and energetic nature.

7. BIRMAN

What a pretty kitty this is. The Birman looks like a less lean, fluffier Siamese, and has a luxuriant creamy fawn-colored coat with dark face and tail. Its legs are dark, too, but the paws are pure white, and it looks as though it's wearing mittens. With angelic round blue eyes gazing out of its dramatic dark face, Birmans are fabulous lookers, blessed with an even-tempered, outgoing personality, and have been described "like puppy dogs in cats' bodies" because of the way they respond eagerly to their owners.

8. American Shorthair

Descendants of the ratting cats that arrived in America in 1620, on board the *Mayflower*, the American Shorthair is an upmarket variety of the non-pedigree domestic house cat. A friendly family-lover, good with children, it's a robust breed that's strong, bright and long-lived. The breed is recognized in more than 80 different colors and patterns, but all share the muscular body, short dense coat and rounded face of the typical house cat. First shown in the late 19th century, the breed was eclipsed for a while by the more glamorous breeds, but made a comeback in the 1960s, when its name was changed from the Domestic Shorthair to the American Shorthair to help it stand out more as a breed in its own right in the show ring.

9. The Oriental

The "greyhound of cats," this svelte feline is like a Siamese, without the typical markings. In their early days, Siamese cats had a range of coat colors, but in the 1920s, breeders decided to develop the pointed pattern that we associate with the breed today. Thirty years later, cat fanciers decided it was time for a bit of variety, and wanted to bring back the red, tabby, black, brown, cinnamon, striped, spotted, mackerel and ticked – to mention but a few color options – Siamese, but they needed to find a new name for the breed to avoid confusion. In physique and personality, the Oriental is just like its Siamese cousin.

10. The Sphynx

Love 'em or hate 'em, the nearly-naked Sphynx is a definite oddity in the feline world. Descended from a mutant hairless kitten born to a black and white cat named Elizabeth in Toronto in 1966, the Sphynx nearly died out for a while, but European breeding programs strengthened the breed, although it is still hard to come by. The cat looks odd in pictures, but aficionados say it has a sensitive and affectionate nature. The Sphynx has the added advantage of being a suitable pet for people who are allergic to cat fur, as it boasts only the lightest covering of down on its skin.

❧ HOW DID CATS SPREAD AROUND THE WORLD? ❧

5000 B.C. Remains of the African wild cat, *Felis libyca*, found in a Cyprus tomb dated at around 5000 B.C. show that at least one cat had crossed the seas from North Africa by this date.

300 B.C. Cats get a mention in both the classic Indian texts *Mahabharata* and *Ramayana*, and were probably brought from Africa by Phoenician traders.

2000 B.C. to A.D. 400 It's impossible to pinpoint a date when cats first set paw in China and South East Asia. But it's a different story in Japan, where traditionally cats arrived in …

A.D. 999 … on the 10th day of the fifth month, to be precise. The real date is probably much earlier, given the closeness of Japan to China. But let's stick with the legend, which has a certain charm. Allegedly Emperor Ichijo was given a pregnant white cat by a Chinese mandarin. The white kittens that arrived were taken as symbols of good fortune for Japan, where white cats still stand for luck and prosperity.

A.D. 100. Hurray! Cats reach Europe, although they would have been here far earlier if the Egyptians hadn't wanted to keep their sacred animals to themselves, and forbidden exports. The rise of Christianity, and the spread of the Holy Roman Empire brought cats surging up through Italy, France and Germany, eventually turning up in Norway and Russia. They also accompanied the Romans across the English Channel at about this time, and a cat's skeleton was excavated from a Roman villa in southeastern England.

1500s. After a long, long gap, cats took another giant leap in their journey around the world, when they were taken by French Jesuits to Quebec in Canada.

1620. A Pilgrim cat or two sailed the ocean blue on board the *Mayflower* in 1620, but cats didn't stake much of a claim to America for another hundred years, when settlers brought them in droves to cope with an explosion of rats.

19th century. Cats completed their circuit of the world when they were exported from Britain to Australia.

*If animals could speak the dog would
be a blundering outspoken fellow, but
the cat would have the rare grace of
never saying a word too much.*

Mark Twain, American author of
The Adventures of Huckleberry Finn

WHICH OF NELSON'S COMMANDERS LOVED CATS?

Admiral Sir John Jervis (1735–1823) was one of Admiral Horatio Nelson's much-loved commanders, and fought alongside him in many great battles. He was named an earl after the Battle of St. Vincent, and when, later in his career, he was replaced as commander in chief of the Mediterranean, Nelson wrote, imploring him to stay:

My dear Lord,

… for the sake of our country, do not quit us at this serious moment. I wish not to detract from the merit of whoever may be your successor; but it must take a length of time to be in any manner a St. Vincent. We look up to you … be again our St. Vincent, and we shall be happy.

The much-missed admiral was born at Meaford Hall in Staffordshire, England, and these touching words, on the tombstone of a cat, are attributed to him:

> *'Tis false that all of pussy's race*
> *Regard not person but the place,*
> *For here lies one, who, could she tell*
> *Her stories by some magic spell*
> *Would from the quitted barn and grove,*
> *Her sporting haunts, to show her love*
> *At sound of footsteps absent long*
> *Of those she soothed with purring song,*
> *Leap to their arms in fond embrace,*
> *For love of them, and not for place.*

❧ IS IT WILD OR TAME? ❧

With its sumptuously spotted coat, the ocicat looks like a miniature leopard or ocelot, and is bred for its savagely good looks. A graceful, athletic cat, with a broad, wedge-shaped face, penetrating golden eyes and an elegant, arching neck, the ocicat originated from a single spotted kitten in a litter produced by crossing a Siamese with an Abyssinian. It's a chatty cat, who loves human companionship.

❧ WHO'S AFRAID OF A HARMLESS CAT? ❧

Ailurophobia, or fear of cats, is not so unusual, and manifests itself in different ways. Some people can't be under the same roof as a cat, others can look as long as they don't have to touch, and still others have a horror of Mr. Baggins suddenly springing on to their lap. Mr. Baggins, of course, knows exactly who feels like this, and acts accordingly.

Through history, people have tormented or even tortured cats. Some had the excuse of being petrified of them, while others were just downright nasty. Brace yourself.

A Succession of Popes

In the 14th to 16th centuries, a succession of popes took to heart the notion of black cats being linked to Satan. Gregory IX issued a papal bull in 1233, allowing the torture, burning and general persecution of cats. Innocent VII upped the ante, and millions more cats got the chop because of him. Innocent VIII reminded the people that a witch's cat should be burned alongside its mistress.

Henri III

The 16th-century king of France passed out cold if he so much as glimpsed *un chat*. Thousands were executed during his reign.

Elizabeth I

Bad Queen Bess's coronation celebrations featured a basketwork effigy of the pope, crammed with cats, which was trundled through the streets, then torched. 'Twas the fashion of the day to persecute cats, but even so …

King Louis XIV

In childhood, the 17th-century French monarch attended a fête where cats were burned alive on pyres. Sorry, this is all rather unpleasant. At least he grew wise enough in later life to acknowledge and regret this hideous act.

Napoleon Bonaparte

Picture the great and noble French emperor, who could look a marauding army in the face without flinching, reduced to a gibbering wreck when a kitten crept into his bedroom. He slashed away with his sword at the tapestry where the bewildered kitten had fled at the sound of Napoleon's imperial screeching.

Johannes Brahms

It's hard to separate fact from pure libel here. The conservative romantic composer allegedly used to shoot cats from his window with a bow given him by the Czech composer Antonin Dvorak. Richard Wagner went further, and claimed that Brahms would reel the cats into his room, and jot down their dying caterwauls, which he would later work into his music. Sound like sour grapes? Quite possibly, as the two were rivals, and there was raging controversy between Brahms' fans and the supporters of Wagner's "neo-German" style of music.

HOW KEEN IS THE BRITISH ROYAL FAMILY ON CATS?

THE QUEEN

Ask Elizabeth II "dogs or cats?" and there's no prize for guessing the answer. Just think Corgis. In fact, none of Her Majesty's immediate family is keen on cats.

PRINCE HARRY

Prince Harry got himself into trouble by hurtling around Allenby Barracks in his Audi A3 coupe and narrowly missing one of the army cats. A witness said, "There was this almighty screech of brakes and I looked up to see his car grind to a halt feet from this startled tabby in the road."

PRINCE ANDREW

Prince Andrew is reduced to tears by the proximity of a cat – not because he's so moved by their feline beauty, but because he's horribly allergic. In the words of what newspapers describe as someone close to the prince, "he hates mogs."

PRINCESS MICHAEL

Princess Michael, wife of the Duke of Kent, is the one royal who is feline-friendly. She has four cats – Siamese and Burmese – which live in regal style, and travel with her between her London and country homes. On a trip to Turkey, the princess was very taken with the Van breed of "fishing cat," and watched these beautiful creatures dive from the fishing boats and come up with their long coats dripping and clinging to their slender bodies, with their catch wriggling in their jaws. She was tempted to take one home, but decided the fish in her pond wouldn't last long if she did.

What happens if you cross a cat with a parrot?
A carrot!

·❀ WHO INSPIRED CROOKSHANKS? ❀·

Crookshanks, as *Harry Potter* fans will know, is Hermione's pet, a large ginger cat with a squashed face and bottle-brush tail, bought in Diagon Alley. Highly intelligent, Crookshanks knew at once that Ron's pet rat Scabbers was actually an "animagus" – a wizard or witch in animal's form. Although Ron initially hated Crookshanks because of his habit of pouncing on Scabbers, later, when Ron acquired Pigwigeon, he used Crookshanks to test that the bird really was an owl, and not another animagus.

Crookshanks isn't a pure-bred cat, since nothing's that simple in *Harry Potter*! He's actually half-Kneazle. And, in case you didn't know, Kneazles are small and cat-like, with oversize ears and a lion's tail. Usefully, the Kneazle can sniff out dodgy characters, hates deception and lies, and has a fabulous sense of direction, so that it can always bring its owners home safely, no matter how lost they are.

J. K. Rowling herself, creator of *Harry Potter* and Crookshanks, isn't keen on cats. She's allergic to them, and prefers dogs. But she formed a distant attachment to a large, fluffy ginger cat she used to see when she ate her lunch in a sunny London square in the late 1980s. He prowled among the humans, but wouldn't be stroked – not that Rowling wanted to get close enough to give herself an asthma attack. She grew fond of this cat's squashed features and used it as her inspiration for Crookshanks – with the addition of bandy legs.

◦◦ HOW DO CATS SURVIVE FALLS? ◦◦

If you've ever let your cat slip from your grasp, or seen her launch for the top of a fence and miss, you'll have seen the righting reflex in action.

A cat who feels herself falling swiftly flips right-side up to avoid landing on her back. Her soft, spongy paws and super-flexible joints absorb the impact when she hits the ground, and give her that nine-life ability to stroll away from even long drops, unscathed.

Two New York vets who studied cats that had made falls ranging from two to 32 stories, found that 90 percent survived. Oddly, falls from between five and 10 floors up were the most likely to cause injuries or prove fatal, and the death and injury rate actually declined for falls from greater heights.

Computer modeling revealed that a cat in free-fall reaches its greatest possible velocity, 60 mph, by the time it has fallen about seven stories. At this stage, though, the cat is still tensed, and less able to cope with the impact. A cat that falls further has time to right itself and relax with its paws spread and limbs flexed, ready for landing. Stretching out the legs like a skydiver helps the cat decelerate, and means its body can absorb the shock through all four paws.

So, next time your cat calls from the top of a tree, don't summon the fire department. Just yell, "It's OK to jump."

Why did the cat run away from the tree?
It was frightened by the bark.

◦◦ HOW DO CATS PURR? ◦◦

Purring isn't easy – try it. Admittedly, the human voice box isn't constructed in the same way as a feline one, but to produce a convincing – all right, fairly convincing – semblance of a purr takes practice, and involves rolling the air around at the very back of the tongue.

Cat anatomy experts have worked out that the purr erupts when air passes across the vestibular folds or false vocal cords at the back of the cat's throat. But how puss manages it is another question, and one that has the experts stumped.

What they do know, is that the anatomy of the domestic cat's throat is different from that of big cats – which is why lions say "RRRRRRRRRAAAAAAAH," rather than "Meow." Lions can't purr, either, even when full of wildebeest. The tiger can give a kind of purr, if you'd care to hang around long enough to give him the pleasure of demonstrating, but he can only do it on the out-breath, so can't achieve that non-stop rumble of delight that comes so naturally to the domestic cat.

⦿ IT'S RAINING CATS AND DOGS ⦿

Cats and torrential rain go together in mythology. Witches turn into cats to ride through storms. Sailors believed cats could predict the weather, and the friskier the cat, the worse the storm on the horizon. "The cat has a gale of wind in her tail." Dogs were guardians of Odin, god of thunder and storms. So it's not surprising that when the cats and dogs are put together, the result is a spectacular storm.

⦿ THE CAT AND MOUSE ACT ⦿

The Cat and Mouse Act was the name given to the British Prisoners (Temporary Discharge for Ill Health) Act, 1913. The aim of this law was to prevent suffragettes from becoming martyrs by starving themselves to death in prison. They could be released as their health deteriorated – and re-arrested at any time.

❧ HOW TO INTRODUCE A NEW CAT TO THE HOUSEHOLD ❧

Do it at a quiet time – not Christmas.

You can bring New Cat home and let it meet Old Cat right away and hope for the best, or you can do some preparation. Keep the cats separate at first. Introduce New Cat's scent by stroking him, then stroking Old Cat to mix the scents. Rub New Cat gently with a cloth, concentrating on his head, where there are most scent glands, then dab the cloth around the house at cat level.

If you can borrow a kittening pen or dog crate, you can keep New Cat in this to start with, so that he and Old Cat can meet safely.

Don't expect miracles. It's an achievement if you reach a stage where the pair will tolerate being in the same room. Anything better than that could take months or even years.

What did the cat say
when he lost his wallet?
I'm paw!

❧ DID DICK WHITTINGTON REALLY HAVE A CAT? ❧

You know the story. Poor boy makes good in the big bad city when he sends his only friend, a cat, to Barbary at the behest of his master, a rich merchant. Dick flees from the house, but drags himself back when the Bow Bells call out, "Turn again, Whittington, thrice mayor of London."

A good thing, too, since he's been sent a small fortune by the king of Barbary in gratitude, because Dick's beloved feline chum has made such short work of the Barbarian rats. Rich Dick marries his master's daughter, Alice, becomes mayor three times as the bells foretold, and, naturally, lives happily ever after. And it's all because of that resourceful cat.

Dick, or Richard as perhaps we should call him, Whittington was a real person who lived 1358–1423. His origins weren't that humble, as his dad was a Sir rather than a servant, but he did become lord mayor three times, marry an Alice and become exceedingly wealthy. But did he have a cat?

Whether there was a household pet lounging around the lord mayor's apartments isn't known. Possibly Whittington, a merchant, made his money trading coal, which came to London in barges known as "cats." Or maybe the idea emerged from the French term "achat" or "purchase," a 14th-century trading term, no doubt frequently on the lips of this canny medieval import/export man.

Or maybe it's simply a story that celebrates the power of the feline hunting instinct. Turn again, Whittington!

❧ A CAT IS FOR LIFE … ❧
"WASH MY BOWL PROPERLY – OR I CALL THE POLICE"

New Zealand cats have had plenty to purr about since the government released the Companion Cats Code, which sets down best practice cat-care rules.

Making sure puss has a cozy bed, a full food bowl – daily for adults, and twice daily for kittens – fresh water, and regularly cleaned bowls and litter boxes reflects the fact that owning a cat is a commitment for the whole of her life, not just while she's a cute kitten.

With a population of between 900,000 and 1.5 million domestic cats, New Zealand has one of the highest numbers of cat lovers in the world. The Code was put together by animal welfare specialists, and parts of it are even enshrined in law, so that it's now a legal offense not to take an obviously sick or scrawny cat to the vet.

It's a sad fact of life that some owners don't do the best they could for their feline charges, but these new rules should give some mistreated cats a better deal. These are some of the minimum standards that New Zealand pet owners must measure up to :

- Dinner should be a high-quality balanced food that's mostly made from meat.
- Cats must have continuous access to fresh, clean water.
- A thin or sick cat must be taken to the vet.
- Food bowls must be spotless and litter boxes regularly cleaned out.
- Kittens must stay with their mother until they're at least eight weeks old.
- A traveling cat must be taken in an appropriate container.

It's all very well intentioned, but in truth most cared-for cats are highly aware of their rights and have no qualms about enforcing them. Why else would there be a fridge magnet that reads:

"Dogs have owners – cats have staff"?

·:· WHAT WAS THE MOST EXPENSIVE PET FOOD COMMERCIAL EVER MADE? ·:·

Could cats be shaking off their image as pets of the lonesome old lady, and coming to the fore as ideal housemates for successful, independent women, looking for someone low-maintenance but cuddlesome to come home to?

A three-part advertising campaign for Sheba cat food, made in the United Kingdom in 2006, had a budget of £1 million, making it the costliest pet food campaign ever.

Angling for a stake in a pet food market worth £7 million a year, a Sheba spokesperson revealed that the ads, which have an "Alice in Wonderland" theme, targeted upmarket women customers who believe in "free-spirited feline mysticism." The advertising company hired Martha Fiennes, sister of actor Ralph, to direct the ads, which starred a ballerina, two top models – and, of course, a British Blue Shorthair cat.

·:· HOW DO CATS KEEP THEIR BALANCE? ·:·

Ears and tail make a great double act that helps Sebastian keep his footing on even the most slender ledge.

Deep inside his inner ear, the organ of balance registers every change of direction and speed, and programs Seb to make the minute adjustments that compensate and prevent a fall.

That remarkable feline skeleton is also put together for easy balancing. The shoulder blades set on either side of the rounded chest, coupled with the vestigial collarbone that's attached simply by muscle, mean he can line up his front paws closely beneath his body to tightrope walk along a fence with poise.

What do cat actors declaim on stage?
"Tabby ... or not tabby?
That is the question."

❧ SO TELL ME WHAT YOU WANT ... ☙

To the doting cat communicator, there are three types of sound:

SOFT MURMURS

Purring and the charming little "prrp" that Fur Boots makes when you give her a passing stroke, mention her name, or simply enter the room.

MEOWS

The vowel-based sounds can be strident, repetitive, demanding, enquiring or complaining. Their precise meaning isn't always that easy to fathom. Vet books say optimistically, "these sounds have subtle variations ... with which owners soon become familiar." Owners may beg to differ, as they go through the morning guessing game ritual:

Fur Boots: Meow! Meow!
Doting owner: Are you hungry? No, the bowl's full.
FB: Meoooow!
DO: You want the door opened?... [FB stares blankly at open door and doesn't move.] Right, not that then ...
FB: Mrrrrow! Mrrrow!
DO: Have you got a pain? Let's have a look. [Picks up cat.]
FB: Prrrrrrr!
DO: You just wanted a cuddle! Why didn't you say so?

SHRIEKS, WAILS AND HISSES

Fur Boots will let rip with one of these if you accidentally trample on her tail, she gets in a fight or, if not neutered, she feels lusty. Cats can come out with a surprising number of grumbles, rumbles, growls, snarls, spits and hisses. It's not a pretty sight, as the whole shape of the mouth changes to an open snarl when a cat is in pain, scared or feeling aggressive, giving a sudden glimpse of the wildcat within.

☙ WHAT WAS THAT DREADFUL NOISE? ❧

As senior cats – we're talking age 14-plus – gradually become more frail, they can begin to feel their loss of independence, and start to need more reassurance that their owners are around and that all is well.

Once Bella has passed her 14th birthday, or thereabouts, don't be alarmed if she sometimes lets fly with a blood-curdling yowl, for no apparent reason. Although it will sound as if she's in agony, this distinctive and penetrating "mrrowl" doesn't mean anything's the matter, and chances are the vet won't find anything wrong with her. She simply wants to make sure that you're home, and she's safe.

What's worst is when Bella gives vent to her terrifying new sound in the depths of the night, and drags you out of bed to reassure her that she's safe and that it's OK to get back to her stress-free slumber. And they say cats are less trouble than babies.

☙ WHAT ARE THE MOST POPULAR NAMES ... ❧

… for cats – and children? It used to be easy to tell the cats from the kids, but not any more. A name like Tibbles, Fluffy or plain old Puss is enough to mark your cat as being past it.

New statistics show that pets are being given the trendiest baby names. These days, if you shout "Molly" out the back door, you're just as likely to bring half the local cats running, as call your daughter in.

SPOT THE DIFFERENCE

We've listed the top names for girls, boys – and cats, in order of popularity. All you have to do is spot the cat in each list. Was it:

1. a) Jessica, b) Jack, c) Tigger (OK, this one's easy!)
2. a) Oscar, b) Emily, c) Joshua
3. a) Max, b) Thomas, c) Sophie
4. a) Olivia, b) Charlie, c) James
5. a) Oliver, b) Max, c) Chloe
6. a) Ellie, b) Daniel, c) Harry
7. a) Chloe, b) Grace, c) Samuel
8. a) Sophie, b) William, c) Lucy
9. a) Harry, b) Molly, c) Charlotte
10. a) Katie, b) Joseph, c) Alfie

11. a) Benjamin, b) Simba, c) Ella
12. a) Charlie, b) George, c) Megan
13. a) Bill, b) Hannah, c) Luke
14. a) Amelia, b) Matthew, c) Jack

Answers – the most popular cat names were
1. c) Tigger, **2.** a) Oscar, **3.** a) Max, **4.** b) Charlie,
5. b) Max, **6.** c) Harry, **7.** a) Chloe, **8.** a) Sophie,
9. b) Molly, **10.** c) Alfie, **11.** b) Simba, **12.** b) George,
13. a) Bill, **14.** c) Jack

Children's names from the British Office for National Statistics, 2005.
Cat names from Norwich Union.

THE WORST CAT NAMES EVER?

Duct Tape	Mr. Fifi	Ikkle
Spayed	Pig	Chicken Chow Meow
Small Man in a Catsuit	Dog Food	Saint Pawl
Sir Meowington Pudger	Bad News	Hannibal Lickter
Cat the Third	Norman Tinkle-Winkle	Captain Beauregard
Bobbobmicbobob	The Urinator	Schmoo-Diddeley
Diphtheria	Stinky	Neuteronomy
Hemorrhoid	eBay	Wisker Wuv

HOW CAN I TELL IF MY CAT'S TOO FAT?

It's as easy for a cat to quietly pile on the pounds as it is for a human, but it's harder to spot the problem because cats don't wear waistbands. Fat cats start by developing a plump paunch, rather than putting on obvious rolls of fat all over like people.

A tubby cat risks health problems such as heart disease, diabetes, cystitis, and arthritis. Your vet can tell you how much puss should weigh. To check her weight, weigh yourself on the bathroom scales, then pick up your pet and weigh the pair of you. Subtract your weight to find out how much she weighs.

You can also get a good idea of whether there's a problem by giving a physical check:

- Stroke your pet's chest, and press gently with your fingertips. You should be able to feel the ribs.
- Look down on your cat and check that the sides of her body taper before the hips.
- Does your pet have a fat, wobbly tummy?
- Gently press on the middle of your cat's spine. Can you feel those bony vertebrae?
- Now stroke her rump and see if you can feel the slight protrusion of her hip bones.
- Does her coat seem poorly groomed around the lower back? She could be too plump to bend around and reach.
- Does she move ponderously, or seem generally lethargic and unenergetic?

If you reckon puss has a weight problem, here's how to tackle it. The root cause is almost certainly this – your cat is simply eating too much.

- You haven't checked the label on the pet food packaging that tells you how much to feed. Many brands carry a table, showing guideline amounts for different weights of cat. Measure out a portion on your kitchen scales, so you can get a real idea of what it looks like. It may well be less than you think, but that's the amount to give daily – not per meal! Many cats don't need a huge amount of food, especially if they're inactive.
- Some owners buy a complete dry cat food, rather than ordinary cat biscuits, but still mix it with wet food. Don't. "Complete" food means what it says. Add anything else, and you're overfeeding.
- Own up – are you slipping puss too many high-fat treats – cream or full-fat milk, fatty meats, chicken skin? Cut them down, or out.
- Check with the family to see just who is feeding puss. Crafty cats soon learn who's a soft touch, and will try their luck with anyone if it means an extra helping.

- Cat snacks can have a place on you cat's menu, but as it is for humans, so it is for cats. Too many of the feline equivalent of crisps and chocolate are bad. Ration treats.
- Someone else is feeding your cat as well as you. The name of the children's book cat-character, Six Dinner Sid, says it all. If you know who to suspect, tell them that your Sid is now on a "special" diet, and mustn't have extra food.
- Your pet has found a local dumpster, where the bags aren't well secured.
- Puss thoughtfully cleans up the leftovers belonging to another household cat or dog.
- An overweight cat becomes lazy, and does less exercise, so it puts on more weight. It's time well spent to devote 10 minutes a day to playing with your pet, tweaking a string for him to chase, encouraging him to leap after a toy, or chase a ball around the carpet.

❖ CAT BACK FACTS ❖

- The feline spine is super-flexible and has pliable muscles, so that Micky is just as comfy curled in a circle, nose-under-tail, or stretched out in a long, languorous line.
- Sinuous as a snake, a cat can twist through 180 degrees in mid-air.
- It takes a drop of a mere 23 inches for Micky to flip himself over, if he should be unlucky enough to take a tumble or be dropped.
- So different from the human skeleton, where movement is limited by the way one bone slots into another, the cat's collarbone is a slender slice of bone which, rather than rigidly connecting the shoulder to the breastbone, floats between the two, held in place by muscle. This clever bit of internal engineering gives Micky a longer stride and a wider range of movement.

⚜ SPLENDIDLY ORIENTAL ⚜

The complete opposite of the rounded and cuddlesome domestic puss, the Orientals would be perfectly at home slinking along a low-hanging branch in some dark and steamy jungle. Descended from Siamese, Korat and Persian felines, they'd fit in just as well patrolling some exotic palace or courtyard. Their sleek, glossy coats, svelte bodies and triangular wedge-shaped faces, give them that haughty, off-with-his-head demeanor.

By nature, they're talkative extroverts, athletic and adventurous – not the pet to choose if you're looking forward to long hours of restful companionship, but perfect if you'd like a cat who'll make it her personal mission to make sure you're never bored.

An ordinary kitten will ask more questions than any five-year-old.
Carl Van Vechten, American writer and photographer

⚜ WAS CHARLOTTE BRONTË FOND OF CATS? ⚜

Indeed she was. According to the biography written by Mrs. Gaskell, Charlotte Brontë (1816–1855), reclusive novelist and author of the melodramatic *Jane Eyre*, found it almost easier to form relationships with animals than with people.

Certainly Charlotte treasured her cat, Tiger, and spoke wistfully of the cat and Keeper, her sister Emily's dog, when she was separated from them while studying in Belgium. She wrote to Emily:

December 1, 1843.

"This is Sunday morning. They are at their idolatrous messe, and I am here, that is in the Refectoire. *I should like uncommonly to be in the dining room at home, or in the kitchen, or in the back kitchen. I should like even to be cutting up the hash, with the clerk and some register people at the other table, and you standing by, watching that I put enough flour, not too much pepper, and, above all, that I save the best pieces of the leg of mutton for Tiger and Keeper, the first of which personages would be jumping about the dish and carving knife, and the latter standing like a devouring flame on the kitchen floor."*

Charlotte returned to her family and pets at Haworth Parsonage early in January 1844, but was soon sorrowing, as Mrs. Gaskell reports:

Soon after she came back to Haworth, in a letter to one of the household in which she had been staying, there occurs this passage. *"Our poor little cat has been ill two days, and is just dead. It is piteous to see even an animal lying lifeless. Emily is sorry."*

Emily Brontë (1818–1848), author of *Wuthering Heights,* had a far more passionate nature than Charlotte. Both sisters loved animals, but their personalities led them to express their feelings differently, as Mrs. Gaskell explained:

"These few words relate to points in the characters of the two sisters, which I must dwell upon a little. Charlotte was more than commonly tender in her treatment of all dumb creatures, and they, with that fine instinct so often noticed, were invariably attracted towards her. The deep and exaggerated consciousness of her personal defects – the constitutional absence of hope, which made her slow to trust in human affection, and consequently slow to respond to any manifestation of it – made her manner shy and constrained to men and women, and even to children. We have seen something of this trembling distrust of her own capability of inspiring affection, in the grateful surprise she expresses at the regret felt by her Belgian pupils at her departure. But not merely were her actions kind, her words and tones were ever gentle and caressing, towards animals; and she quickly noticed the least want of care or tenderness on the part of others towards any poor brute creature. . . .

"The feeling, which in Charlotte partook of something of the nature of an affection, was, with Emily, more of a passion. Someone speaking of her to me, in a careless kind of strength of expression, said, 'she never showed regard to any human creature; all her love was reserved for animals.' The helplessness of an animal was its passport to Charlotte's heart; the fierce, wild intractability of its nature was what often recommended it to Emily."

Charlotte Brontë included cats in her novels, and in *Shirley* it is Robert Moore's kindness to animals that attracts the compassionate Caroline Helstone to him.

"I know somebody to whose knee the black cat loves to climb, against whose shoulder and cheek it likes to purr. The old dog always comes out of his kennel and wags his tail, and whines affectionately when somebody passes. He quietly strokes the cat, and lets her sit while he conveniently can; and when he must disturb her by rising, he puts her softly down, and never flings her from him roughly: he always whistles to the dog, and gives him a caress."

❧ HOW ARE CATS LIKE TEENAGERS? ❧

Is it the way they wander off when they're bored, leaving their belongings – the catnip mouse, the length of string – scattered around on the floor?

Or the way they want FOOD, on demand, but only the food they like, not something you cooked specially for them?

The way they tend to ignore you – until they're after something?

Or their habit of treating you like a servant? Where a cat meows for you to open a door that's already ajar, a teenager leaves his cereal bowl on the table, rather than walk a few feet to the dishwasher.

Most of all, maybe it's the way you continue to love them fiercely, despite their little foibles.

❧ WHICH CAT JOINED A JAZZ CLUB? ❧

Olsen, a plump Siamese chocolate point and pet of the Bishop of Chester, found life on the grounds around Chester Cathedral rather dull, once he'd sampled the excitements of getting locked in the belltower. He took to roaming off into the old city at night, coming back later, and later, and later, until early one morning, he returned home with a note tucked into his collar from the nearby Alexander's Jazz Theater – a bill for entrance fee, drinks and food. Olsen had joined a jazz club.

It turned out Olsen was a regular at the club and seemed to enjoy the music so much that the tenor sax player was tempted to take him home. The bill was a way of tracking down Olsen's owners. The bishop went to the club, whose owners assured him that Olsen was a welcome visitor, and would continue to get prawns on the house.

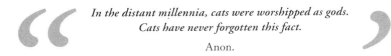

In the distant millennia, cats were worshipped as gods.
Cats have never forgotten this fact.

Anon.

❧ HOW TOUCH-SENSITIVE IS MY CAT? ❧

In a word – very. A cat's body is a veritable hotbed of nerve endings, which is why they sooooo love to be stroked and caressed. Some even enjoy being washed by another cat. It just feels delicious.

CAT NURSERY RHYMES

Hey diddle, diddle,
The cat and the fiddle,
The cow jumped over the moon;
The little dog laughed to see such fun,
And the dish ran away with the spoon.

This traditional nursery rhyme first appeared in print in 1765, under the title "High Diddle Diddle." There are lots of variations on this rhyme – the dog can laugh at "fun" or "sport," and sometimes the last three lines go:

The little dog laughed,
To see such craft,
While the dish ran after the spoon.

"Craft" may have referred to witchcraft. There are many explanations of the rhyme to do with royalty, politics and even Egyptian mythology, but most seem pretty contrived, and it was probably written as a simple nonsense rhyme to amuse small children.

A cat came fiddling out of a barn,
With a pair of bagpipes under her arm;
She could sing nothing but fiddle-cum-fee,
The mouse has married the humble-bee;
Pipe, cat, dance, mouse,
We'll have a wedding at our good house.

We are all in the dumps,
For diamonds are trumps;
The kittens are gone to St Paul's.
The babies are bit,
The Moon's in a fit,
And the houses are built without walls

Fee-dum, fiddle-dum fee,
The cat's got into the tree.
Pussy, come down,
Or I'll crack your crown,
And toss you into the sea.

Kings are often symbolized by lions (big cats), so maybe this rhyme refers to the episode in 1651 when King Charles II was defeated at the Battle of Worcester and hid in an oak tree to escape from the Roundheads. His crown was already cracked, as he was recognized as king only in Scotland, and he was forced across the sea to exile on the Continent before being restored to the throne in 1660.

Ding dong bell,
Pussy's in the well.
Who put her in?
Little Tommy Green.
Who pulled her out?
Little Tommy Trout.
What a naughty boy was that
Who tried to drown poor Pussy Cat
Who never did him any harm,
But killed all the mice in Father's barn.

Traditional nursery rhyme. In the original, the cat was left to drown, but later the words were amended to teach children that it's wrong to harm an animal. The boys' names appear in different versions as Tommy Thin or Tommy Lin and Tommy Stout.

Pussy cat, Pussy cat,
Where have you been?
I've been up to London
To look at the Queen.
Pussy cat, pussy cat,
What did you there?
I frightened a little mouse
Under her chair.

There's a story that one of the courtiers of Queen Elizabeth I kept a cat, which wandered under the queen's throne and made her jump when it brushed her foot. Fortunately the queen saw the funny side of it and, far from being sent to the Tower of London, the cat was given a royal pardon.

When I play with my cat, who knows whether she isn't amusing herself with me more than I am with her?

Michel de Montaigne (1533–1592)

WHO'S KING OF THE CATS? A GHOSTLY TALE

All was snug in the cottage, and the Missis and the old tom cat were basking by the fire, though outside all was black and a storm was raging. Suddenly, the door flew open and a man pitched himself in and slammed the door behind him. His wife leaped to her feet, and even the old tom cat lifted his head.

"What's on Earth's the matter?" asked Missis.

"I was on the road to home, all alone, with the wind raging around me," said the man, "and I saw a sight would make your hair stand on end. I came across a procession of cats."

And as he said this, his cat quietly uncurled itself and came towards him, looking as though it were listening intently.

"And at the front were four cats, and they were bearing – a coffin!"

At this, the man's own cat fixed his huge, green eyes on the speaker's face, as if willing him to go on.

"On top of the coffin was a small cushion – and on that cushion lay a crown!"

And suddenly their old black tom swelled up to twice his size and hissed out these words –

"So! Old Tom's dead and I'm King of the Cats!"

And he turned around three times and flew up the chimney before either of them could stop him.

From an old Irish folk tale.

❧ HOW CAN I TEMPT A FUSSY CAT? ❧

Are you sure you want to? The reality is that some cats are arch-manipulators, and use every trick in the book to persuade you to give them only the type of food that they're hooked on. If Tom-Tom gets hungry enough, chances are he will eat the less favored food. But why should he worry. He knows you'll cave in when he takes one sniff and turns away, sadly.

But before you start trying to retrain him, check that nothing's putting him off legitimately.

- Is the food the right temperature? Cold food just doesn't smell right. Room temp or slightly warmer is better.
- You can't blame a cat for refusing canned food that's been out for a few hours. If the surface looks dried, throw it away, scour the bowl, rinse thoroughly and try again.
- If your cat licks up the gravy or jelly in its food, but leaves the chunks, try mashing the whole dish together, or switching to a meat-only type.

- Is the food really fresh? Canned food can turn within a day. Dry food goes stale unless kept in an airtight container, and can pick up flavors from nearby damp or strongly-scented items. Don't store the litter supply next to the air freshener or washing powder.
- Offer only a small portion. Some cats are put off by a big bowl full.
- It's worth trying raising the bowls a little off the floor, as some cats, such as older ones with creaky joints, can't crouch in comfort.

RETRAINING A FUSSY EATER

- Mix the favorite food in with other food.
- Give treats such as fish or liver only occasionally. Otherwise your cat might start to demand them all the time.
- Nip faddiness in the bud as you wean a kitten, and introduce your cat to a whole range of flavors.
- Buy a mixture of flavors and swap them around so that your cat is used to having different foods and doesn't have a chance to get fixated on one.

❧ WHO PLAYED MRS. NORRIS? ❧

Argus Filch, caretaker of Hogwarts School in the *Harry Potter* series, patrols the corridors with his cat, Mrs. Norris, "a scrawny, dust-colored creature with bulging, lamp-like eyes just like Filch's."

In the film *Harry Potter and the Philosopher's Stone,* the role of Mrs. Norris was shared between three Maine Coon cats, two males and one female, Pebbles.

Pebbles was a pedigree breeding cat, but after she'd had to have caesarean deliveries for both her litters, she had to be spayed. Without kittens to amuse her, Pebbles got bored, and her owner decided to rehome her.

Then Gary Gero, one of the world's leading experts in animal training, heard about Pebbles. He was looking for a Maine Coon to play the part of Mrs. Norris, and came to see Pebbles. She passed the audition effortlessly and Gary gave her a share in the part with two male cats that he'd already found.

The males responded better to training, so Pebbles took center stage in the freer scenes, where Mrs. Norris is seen roaming up and down the corridors of Hogwarts, on the lookout for Peeves the Poltergeist.

❧ THEATER CATS ❧

No self-respecting theater used to be without its resident cat – because what better lair for rats and mice, than the dusty backstage nooks and crannies? A rat-catcher was as much needed as a scene-shifter to keep life behind the curtain running smoothly.

And rather than hitting the gin, a terrified actor could soothe his first-night nerves by fondling a friendly theater cat. Cat lovers, being a superstitious crowd, like the idea of cats being good omens for a show's success. It was even suggested, in *The Curse of Macbeth* by Richard Huggett, that the luckiest charm of all was finding a cat dropping in the dressing room – as discovered by Noel Coward on the opening night of *The Vortex*. The show ran and ran.

But the era of health and safety has put an end to all that, and the legendary lineage of theater cats have taken their last bow.

AMBROSE

A smartly-attired black-and-white cat, Ambrose regularly upstaged Michael Crawford by roaming on to the set of *Billy* when the actor was in full flow.

CHESS

A vast checkerboard of a cat, Chess used to snooze in the box office of the Manchester Apollo and roam around the theater.

BEERBOHM

Beerbohm of the Globe (latterly renamed the Gielgud), a rotund tabby with a white chest and red collar, was given an obituary on the front page of *The Stage* when he died. His oil portrait hung near the dressing rooms, and he would generally light on one cast member for each production, and share his or her dressing room for the duration of the show's run. Beerbohm, named after celebrated actor-manager Herbert Beerbohm Tree, felt it was *de rigueur* to test the cast's improvisation skills, by making an unscripted stage appearance at least once during every production. He eventually retired, and went to live with one of the theater's carpenters.

BOY CAT

Boy Cat and his companion – you've guessed – Girl Cat, were called in to deal with a mouse problem at the Albery Theater. Still hungry, Boy Cat distinguished himself at the gala performance by eating Princess Margaret's bouquet. Bored one night, he ambled across the stage, oblivious to the first act of *Pygmalion* that was going on at the time, hopped down into the auditorium and spent the rest of the show curled up on an empty seat in the front row, being stroked by an audience member.

Gus

Gus the Cat at the Theater Door in T. S. Eliot's *Old Possum's Book of Practical Cats* had aspirations way beyond those of a humble rodent operative. "I have played in my time every possible part."

Misty

Misty lived beneath the stage of the Strand Theater in London, and her bed was a box of old theater curtains.

WHICH CAT WAS CLONED?

Why have just one of your pet – if you could have two? It's now possible to create a carbon copy of Tinkerbell, if one of her simply isn't enough.

In 2003, the Texan owner of Nicky was so distraught when her 17-year-old pet died that she paid $50,000 to have her cloned. Little-Nicky was produced from Late-Nicky's DNA, and was born in 2004. Along with other similarities, he shares his cell-mate's fondness of water.

WHY DO CATS PURR?

Those sonorous vibrations that seem to emanate from the very core of a cat must be one of the most soothing sounds in the world.

Kittens purr before they can even open their eyes, perhaps to reassure their mother (who will probably be purring in reply) that they are alive and well.

That cats purr when they're content, warm, dozing or being stroked by their owner – preferably all at once – is self-evident. But cats also purr when they're in trouble. An injured cat, or one who's in pain, may purr. Even a dying cat might purr, as owners who've accompanied their pet through that last sad trip to the vet will know. Purring seems to be the default sound for a cat, and, like so many things about them, it can have a multitude of meanings.

❖ DOES THIS SOUND FAMILIAR? ❖

This proverb is found in numerous languages – can you guess what it means?

Kdy? není kocour doma, my?i mají pré (Czech).

Als de kat van huis is, dansen de muizen (Dutch).

Kui kassid läinud, on hiirtel pidu (Estonian).

Kun kissa on poissa, hiiret hyppivät pöydällä (Finnish).

Quand le chat n'est pas là, les souris dansent (French).

Ist die Katze aus dem Haus, tanzen die Mäuse auf dem Tisch (German).

Gdy kota nie ma, myszy harcuja (Polish).

Quando o gato sai, os ratos fazem a festa (Portuguese).

Când pisica nu-i acasa, joaca soarecii pe masa (Romanian).

Cuando el gato se ausenta, los ratones bailan (Spanish).

När katten är borta, dansar råttorna på bordet (Swedish).

Yes, it's "While the cat's away, the mice will play" – although depending on their nationality, the mice play, dance, have a festival or reign. Seems it's the same the world over. As soon as the boss's back is turned …

❖ WASHING A CAT ❖

A warm shower is not most cats' idea of fun, but if you're going to shower your pet, or if she's rolled in something truly disgusting or potentially harmful, you've no choice but to give her a bath.

If your cat is going to be bathed regularly, it's best to get her used to the idea in kittenhood. If this is the first time for an adult, call in a friend to help. Otherwise, you'll go crazy trying to keep Flossie still while you lather the shampoo. Get everything ready before you start, have a good supply of towels handy, and keep up a soothing banter to your cat all the time.

1. Put about two inches of warm water into a sink, or stand a bowl in the bath and use that. Use a showerhead on a gentle, warm setting to soak the fur, then rub in a little feline or baby shampoo and lather.
2. Rinse well with the shower, then press and squeeze as much water as you can from the coat. Lift the cat out and towel her gently.

3. You can try using a hairdryer on the lowest setting to dry your cat, but if she finds that too traumatic, keep on patting and rubbing the coat with towels until she is as near to dry as possible.

4. Finally, brush through the coat to make sure it's lying properly and, *voilà*, one freshly laundered and beautiful feline!

❧ HOW TO GIVE YOUR CAT A HOME HEALTH CHECK ☙

You can do this quite easily when you're stroking or grooming your pet. If you find anything unusual, book your cat for a veterinary check up.

- Eyes should be clear, inner eyelid retracted. The nose should be clean and slightly moist. The ear should be clean with no signs of wax or discharge.
- Draw your hands gently along the cat's body and over her underbelly to feel for swellings or lumps or painful areas.
- If you can persuade Gemma to open her mouth, check her breath and take a look at her teeth and gums for any signs of infection.
- The fur around her back passage should be clean. Staining can be a sign of various health problems.
- Put a comb gently through her coat and shake it on to a piece of white paper. Black specks indicate fleas. Treat immediately to prevent any further infestation.

❧ A CAT OF BEAUTY IS A JOY FOREVER ❧

John Keats (1795–1821), the Romantic poet who once thrashed a butcher boy he caught tormenting a cat, wrote this sonnet about an aged feline in 1818 during a period when he was writing a new poem almost daily.

To Mrs. Reynolds' Cat

Cat! who hast pass'd thy grand climacteric,
How many mice and rats hast in thy days
Destroy'd? – How many tit bits stolen? Gaze
With those bright languid segments green, and prick
Those velvet ears – but pr'ythee do not stick
Thy latent talons in me – and upraise
Thy gentle mew – and tell me all thy frays
Of fish and mice, and rats and tender chick.

Nay, look not down, nor lick thy dainty wrists –
For all the wheezy asthma, – and for all
Thy tail's tip is nick'd off – and though the fists
Of many a maid have given thee many a maul,
Still is that fur as soft as when the lists
In youth thou enter'dst on glass-bottled wall.

Keats died of tuberculosis in Rome when he was only 30, and was buried in the Protestant Cemetery where his gravestone reads: "Here lies one whose name was writ in water."

The picturesque and peaceful burial ground, known to the Italians as *Il Cimitero Straniero* – the Foreigners' Cemetery – is inhabited by countless cats, whose food bill is financed by visitors. Cats of every type and color lounge in the sun, doze on tombtops, dangle themselves over gravestones and stroll along the tops of the ancient stone walls. Some are timid, but others enjoy meeting tourists and will hop eagerly onto a lap if a tourist sits down to rest from the clamor of the city outside the cemetery walls.

What's a cat's favorite TV program?
The 7 o'clock mews!

WHICH CAT WAS AVAILABLE BY MAIL?

The California Spangled cat was created by writer/director Paul Casey, who was perturbed by the dwindling numbers of wild cats in Africa. He wanted to breed a domestic cat that looked like a wild one, in the hopes of discouraging people from buying fur coats – because who would want to walk around wearing something that looked like their pet?

The breeding program took 10 years, and the California Spangled Cat was the result, with its long lean body, and dense satiny coat cloaked with leopard-like spots. It's an exotic mix of Malayan house cat with a scattering of Angora, Abyssinian, Manx and Siamese thrown in.

Inexplicably, Casey decided to launch the breed through the mail-order catalogue of the exclusive Beverly Hills branch of Neiman Marcus. The strategy did raise funds for wildcat protection, but raised hackles elsewhere. Neiman Marcus wasn't keen on the anti-fur-coat strategy, animal rights groups were against the deliberate breeding of domestic cats, and the pedigree lobby thought there were enough spotted breeds already. Oh dear. No wonder the breed has struggled to achieve recognition.

DO CATS APPRECIATE MEXICAN ART?

Bob Walker and his wife were keen collectors of colorful Mexican art. Their house in San Diego was stuffed full of furniture, masks, pottery, frames and carvings – and cats. Nine of them, including Benjamin, Beauregard, TomCat, Jerry Lee, Celeste. Every year, the couple adopted a new kitten from cat rescue. Then Bob constructed a floor-to-ceiling scratching post to give all those claws some sharpening.

The cats approved, and gradually there was more, until the whole house became a network of special feline walkways, steps, balconies, boxes, alcoves, entrances and exits, all of them decorated in brilliant reds, blues, yellow, purple and green. A scarlet cat staircase runs up the side of a daffodil wall and snakes off, disappearing through a cat-sized archway into the next room. A sea-green branch-way takes off across the ceiling, to a roosting perch where cats can dangle above the carefully arranged ornaments below.

What became known as the Cat House is designed so the feline residents can wander from room to room on their own, specially constructed, vibrantly colored walkways, squeezing through specially made holes to get from one room to the next. The whole house is an artifact in itself.

·:❖ WHAT'S TO DRINK? ❖:·

- Although you should always put down a fresh drink for your pet, cats get most of the liquid they need from food, which is why you'll seldom spot them lapping at their water.
- Some cats can't cope with fresh cow's milk, and they certainly don't need it. Give milk in moderation only, when you're sure it won't cause an upset stomach.

> *The Cheshire Cat vanished quite slowly, beginning with the end of the tail, and ending with the grin, which remained some time after the rest of it had gone.*
>
> *From Alice in Wonderland,*
> by Lewis Carroll (1832–1898)

WHAT'S NEW PUSSYCAT?

The plethora of cat-decorated items, and things shaped like cats, knows no end. You can indulge your obsession to your heart's delight – just wander around a gift shop, or browse the Internet, and you'll find cats prancing, lounging or just being cats, on any number of things.

eyeglass holders	T-shirts	paperweights
notebooks	calendars	vases
oven mitts	clocks	pencil holders
teatowels	cookie jars	stationery
tea cozies	cutting boards	napkins
nightlight holder	storage jars	bottle stopper
apron	doormats	tea infusers
computer mouse	doorstops	coasters
mouse pad	duvet covers	table mats
screensaver	teapots	fridge magnets
gloves	trivets	car mats
mittens	trays	earrings
jumpers	cushions	tote bags
scarves	mugs	luggage labels
hats	jewelery	

☙ WHY DID TOLKIEN WRITE ABOUT THE FAT CAT ON THE MAT? ☙

When J.R.R. Tolkien wrote a simple poem called "Cat," he wasn't just jotting down random thoughts about felines. The scrawled letters "SG" in the margin of the page where he wrote the poem indicate that he may have intended to put the poem into the mouth of Sam Gamgee, Frodo's homely Hobbit companion in *The Lord of the Rings*.

What do you do with a blue Burmese?
Try to jolly it along.

☙ WHY IS MY CAT SUCH A FUSSY EATER? ☙

Blame their highly sophisticated senses of taste and smell. Cats have thousands of tastebuds to help them discriminate between fresh foods, and those that are unsafe to eat. Nikita knows if you put down a plate of meat for her, or try to fob her off with the remains of your vegetable stir-fry, because her taste buds are wired up to respond to meaty flavors rather than veggie ones.

Her nose is crammed with twice the number of scent-sensitive cells as yours, so food that smells just fine to you can be rejected as decidedly past its use-by date by her.

A tiny kitten can tell the difference between one of its mother's nipples and another, simply by sniffing. In later life, scent is all-important for detecting feline territory markings, and smelling the difference between friend and foe.

Cats have another scent detector, the Jacobson's organ, built into the roof of the mouth. They use their tongues to lap odors on to this receptor, giving them another way of refining scents and understanding their surroundings.

❧ A SMUGGLER'S TALE ❧

A man who tried to smuggle two leopard-cat kittens into Los Angeles by hiding them in his backpack was fined $50,000.

Chris Mulloy nearly got away with it in the kerfuffle caused by his traveling companion, who was grabbed by customs officers after an exotic bird flew out of his suitcase. He then admitted to having two pygmy monkeys hidden in his trousers.

❧ WHO'S HAVING FUN AT 50? ❧

Dr. Seuss's wondrous creation, *The Cat in the Hat*, has celebrated a half-century of helping children to read – and making it so much fun, they don't realize they're learning!

The original inspiration behind the anarchic cat came from an article in *Life* magazine, "Why Johnny Can't Read," which said that children struggled to read because primers were boring. Ted Geisel, aka Dr. Seuss, was already known as an illustrator, and he rose to the challenge and dreamed up *The Cat in the Hat*. The books feature a larger-than-life character who bounds into children's lives, gives them a dose of crazy fun that would set their parents shrieking, then disappears just as Mother comes home. The text cleverly uses simple, new-reader vocabulary, but in a way that's far more engaging than "the cat sat on the mat."

Ted Geisel died in 1991. His books have sold more than 200 million copies worldwide, and children still love them.

❧ HOW SHOULD I GROOM MY CAT? ❧

Shorthaired cats don't need much grooming, but fluffy kitties definitely do. If your cat's a pedigree, get the breeders to show you how to groom. Otherwise, simply arm yourself with the right kit, cover your lap with a towel, settle Suzuki comfortably, and off you go.

WHAT YOU'LL NEED

- Bristle brush or round-tipped grooming brush
- Wide-toothed metal comb
- Fine-toothed metal comb
- Unscented talcum powder

With a shorthaired cat, simply work your way gently through the coat with a brush, and use a comb to work out any tangles. Rub your cat's coat lightly with a piece of silk, soft cloth or even just with your hand when you've finished, to give the coat a glossy shine.

For longhairs, the whole business is more elaborate. First, seek out any knots and tangles and tease them out gently with a comb. Then brush thoroughly to lift the fur and get rid of loose hairs. Rub a sprinkling of talc through the coat once a week, to help clean it. Remember to brush the tummy and tail as well. Next, work on separate sections and brush each one firmly towards the head and against the natural lie of the fur. Do this carefully, to remove any talc. Finally smooth the coat with the brush, and use the comb to finish the ruff and neck fur.

❧ WHAT HAPPENED TO THE KATRINA CATS? ❧

Cats were just as much victims as humans when Hurricane Katrina hit New Orleans in 2005. The city's feral cats were made homeless when the streets that were their home were flooded. Ten of them were brought to a shelter and because they were so untamed, were housed in Shy Cottage, where 30 other timid cats reside.

The hurricane refugees were each named after a 2005 hurricane – Emily, Jeanne, Harvey, Nate, Philippe. And in their paperwork, a capital "H" next to each name is a reminder of how they came to be rescued.

These cats have never been domesticated, and are too wild to be adopted into a home, so they'll spend the rest of their lives in Shy Cottage – a great improvement on their pre-Katrina hangout, a parking lot behind a New Orleans hairdresser's.

❧ THE OWL AND THE PUSSY CAT ❧

The Owl and the Pussy Cat went to sea
In a beautiful pea-green boat,
They took some honey, and plenty of money
Wrapped up in a five-pound note.
The Owl looked up to the stars above,
And sang to a small guitar,
"O lovely Pussy, O Pussy, my love,
What a beautiful Pussy you are,
You are,
You are!
What a beautiful Pussy you are!"

Pussy said to the Owl, "You elegant fowl!
How charmingly sweet you sing!
O let us be married! too long we have tarried:
But what shall we do for a ring?"
They sailed away, for a year and a day,
To the land where the Bong-tree grows
And there in a wood a Piggy-wig stood
With a ring at the end of his nose,
His nose,

His nose,
With a ring at the end of his nose.

"Dear Pig, are you willing to sell for one shilling
Your ring?" Said the Piggy, "I will."
So they took it away, and were married next day
By the Turkey who lives on the hill.
They dined on mince, and slices of quince,
Which they ate with a runcible spoon;
And hand in hand, on the edge of the sand,
They danced by the light of the moon,
The moon,
The moon,
They danced by the light of the moon.

From "The Owl and the Pussy Cat" by Edward Lear
(1812–1888)

❧ HOW DO CATS MARK THEIR TERRITORY? ❧

Your cat likes nothing better than to mark out your house and garden as his – and his alone.

If you own an un-neutered tom, or are unlucky enough to be visited by one who sneaks in through the cat-flap, then you'll know all about spraying. The unmistakable smell of cat urine is a sure sign that somebody has been staking a claim. Blasting a patch of furniture, wall or garden with pungent pee is the male cat's way of leaving his own trail and warning other feline males to keep off.

Spraying is a highly antisocial way of marking territory, peculiar to toms who haven't undergone the snip. But all cats like to define their home, and their people, with a daubing of scent. Fortunately, these other markers can't be detected by the human nostril.

Rubbing lets puss deposit sebum, a substance produced by sebaceous glands in the hair follicles around a cat's face, lips, chin, eyelids and tail base, anywhere he wants to leave his stamp. That's why you'll often see your cat rubbing around furniture, trees and, especially, your legs. He wants everyone in the cat world to know that you belong to him.

Cats also like to spread their scent over their own body, one reason why they spend so much time licking their bodies. Although the primary purpose of the tongue-bath is to clean the fur, the saliva also carries the cat's individual scent. A thorough going-over means that every inch of your pet's body carries his own delicious aroma, even if it's only perceptible to him, and other cats.

When you catch your cat clawing at the back of the sofa, he's not just doing it to sharpen his claws and wind you up. Sweat glands at the base of the paw release his own familiar *eau-de-puss* and what better than an energetic upholstery-ripping session to make sure that the furniture is well impregnated?

❧ TAIL TALK ❧

Archie's tail has a life of its own and can go from being a smooth coil that keeps its sleeping owner cozy, to an erect and bristling alarm-signal when he scents danger.

Greeting a friend: tail is held up high, the tip bent slightly forward. This position allows another cat to sniff the under-tail area to check whether the scent spells friend or foe. Delicious!

Kittens greet their mothers with their tails held aloft, and will rub mom's rump and around the top of her tail to try and persuade her to give them some food. Adult cats go through a similar performance when they twine their tails around their owners' legs, begging for attention.

Playtime: When cats go crazy they may fluff their tail into a big fierce bottle-brush, and carry it in an inverted U-shape.

Ready for action: A cat who's deciding what to do may swish his tail from side to side while he ponders the best course of action. As Archie gets more excited, his tail swings wider and faster. The result can be a bout of boisterous play with another cat – or a fight. A tail that's thrashing violently means massive excitement, or rage.

Keep off: If there's a serious threat from another cat, or a dog, Archie can stiffen his tail-hairs so they stand out like a bottle-brush. He can double the size of his tail like this, because animals on the defensive need to look as BIG as possible to scare their opponents.

A cat's got her own opinion of human beings. She don't say much, but you can tell enough to make you anxious not to hear the whole of it.

Jerome K. Jerome, English author of the comic novel *Three Men in a Boat*

⊛⊛ WHAT'S IN A WHISKER? ⊛⊛

Those charming whisker are rooted in sensitive pads which can detect even a slight change in air currents. Whiskers don't just grow on the muzzle. They sprout from eyebrows and between the pads of the front paws as well. They're rooted three times deeper than a normal coat-hair, and their follicles are embedded in a sensitive capsule which transmits sensations to a network of highly attuned nerves all around it.

Like the ears, the whiskers are super-mobile and can be tweaked and twiddled to pick up on the slightest nuance. So sensitive are they that they respond if displaced by a distance of an incredible 2,000 times less than the width of a human hair.

Cats use their whiskers to judge the width of tight spaces. If you can't fit your whiskers through, then it's better not to try.

❧ WHO STOLE THOMAS HARDY'S HEART? ❧

Thomas Hardy (1840–1928) wasn't just author of tragic pastoral love tales such as *Tess of the d'Urbervilles, Jude the Obscure* and *Far from the Madding Crowd*. He was also a cat lover.

He had at least eight cats at his house, Max Gate, where saucers of milk were put out on the lawn for other cats who arrived for tea.

His first wife, Emma, shared Hardy's passion, and even when their relationship was in deep decline they could always talk enthusiastically about the pets they loved so much. In Hardy's letters to Emma, he told her how he'd had to reimburse a maid because one of the cats had destroyed her hat, and how he'd asked another servant to cuddle a cat whose kittens had been taken away.

In the garden was a pets' graveyard, complete with headstones. Hardy was so distressed when one cat died, that he wrote a poem about it:

Last Words to a Dumb Friend

Pet was never mourned as you,
Purrer of the spotless hue,
Plumy tail, and wistful gaze
While you humored our queer ways,
Or outshrilled your morning call
Up the stairs and through the hall –
Foot suspended in its fall –
While, expectant, you would stand
Arched, to meet the stroking hand;
Till your way you chose to wend
Yonder, to your tragic end.

Never another pet for me!
Let your place all vacant be;
Better blankness day by day
Than companion torn away.
Better blot each mark he made,
Selfishly escape distress
By contrived forgetfulness,
Than preserve his prints to make
Every morn and eve an ache....

Housemate, I can think you still
Bounding to the window-sill,
Over which I vaguely see
Your small mound beneath the tree …

As an old man, Hardy was given an orange-eyed Persian, Cobby, who stayed by his side until he died. And thereby hangs the curious tale of Hardy's heart.

Hardy's body was destined to rest in Poets' Corner in Westminster Abbey. But because he had wanted to be buried beside Emma, in the Dorset village of Stinsford, it was decided that his heart would be buried there, in the graveyard of St. Michael's Church. What happened next is still something of a mystery. The local doctor removed the heart, and left it overnight wrapped in a tea towel – in a sealed cracker box, say some versions – ready for local burial. When he returned next day, the box was open and all that remained was a bloodied towel and a few fragments of gristle. Someone had eaten Hardy's heart.

From this point, you can write your own ending, from one of the numerous versions of this tale.

- Perhaps the cat was Hardy's very own Cobby who may, or may not, have disappeared after the writer's death.
- Maybe the culprit was another cat belonging to Hardy's housekeeper, sister or even the doctor.
- Was the heart ever placed in a tin, or was it snatched from the kitchen table?
- And what was buried in its place? Was it the heart of a pig, a calf or, in a particularly melodramatic version, did the undertaker strangle the feline thief and bury its body in place of its master's organ?
- Or is the whole tale simply a story worthy of Hardy himself, a rural, rather than an urban, legend with no foundation in truth?

⟡ WHERE DO BURMESE CATS COME FROM? ⟡

A single cat, Wong Mau, was the mother of the Burmese breed, which began in 1930 in the United States. With a glossy mid-brown coat with darker brown points, she was described as "a rather small cat, fine boned, but with a more compact body than that of a Siamese, with shorter tail, a rounded, short-muzzled head, with greater width between rounded eyes."

This solitary puss was brought from Burma to New Orleans by a sailor, intrigued by a creature that looked nothing like any cat he'd ever seen before.

Wong Mau found her way into the hands of Dr. Joseph G. Thompson who was so taken with her that he gathered a group of scientists and cat breeders to try to isolate her distinguishing characteristics and produce a line of Wong Maus as a breed in their own right.

If Wong Mau looked like anything it was a Siamese, so she was mated with a Siamese male. Scientists experimented with breeding from the kittens, until they managed to produce both Wong Mau look-alikes and an even more appealing cat with a solid dark brown coat. Meet the American Burmese.

Described variously as "Velcro cats" because they love to cling and "bricks wrapped in silk" – can't you feel it? – the richer-than-rich cocoa brown coloring reigned supreme for decades, until newer "dilute" coat colors were allowed by the Cat Fanciers' Association, or CFA, the major American cat registry, "dedicated to the preservation of the pedigreed cat."

Now, choosing a Burmese color is like visiting a huge department store. "Which would Madam prefer – original Sable, Champagne (honey brown), Blue (warm fawn undertones), frost, or platinum (pale fawn with lilac frosting)?"

CATS AND COMPUTERS 5
I just need to remind myself
how nasty the wires taste.

●❀ HOW CAN I MAKE MY HOME CAT-SAFE? ❀●

- Keep chemicals such as insecticides, household bleach and antifreeze well sealed and out of reach.

- Arrange electric cables so they're inaccessible when you can. Spray loose wires with cat repellent.

- Don't use products sold for use on dogs to treat your cat. Substances that are safe for canines can be harmful for cats.

- Deter your cat from jumping up on to the stove by sticking some double-sided tape on the edge of nearby worktops. A sticky landing is enough to make most cats say "never again."

- Look inside your washing machine and tumble dryer before turning on. It has been known for pets to make their bed inside.

- Although cats usually give babies and young children a wide berth, put netting over carriage and crib as a precaution.

- Check around and under your car before driving off. Cats love to rest in dark spots and could be in danger.

- All cats should wear a collar with some form of identification attached, or be fitted with a microchip under the skin of the neck, in case they are lost or injured. Even indoor cats need naming as a precaution, so that should they ever get out, they can be returned safely.

✦ WHEN WAS THE FIRST CAT SHOW HELD? ✦

Crystal Palace in London was the venue for the first-ever U.K. cat show, held on July 13, 1871.

The show was set up by author and artist Harrison Weir, who wanted to raise the feline profile so that "the too often despised cat will meet with the attention and kind treatment that every dumb animal should have and ought to receive."

There were 170 entrants, including Angoras, Persians and a Scottish wild cat with a missing paw. The exhibits were caged, and given crimson cushions to sit on. Eastern and other foreign breeds, and native British cats were shown in 25 different classes, including one for "Cats Belonging to Working Men." Exhibiting cats was a pastime of the gentry in those days. The judges included Weir and his brother and their friend the Rev. J. Macdonna, a breeder of St. Bernards. It's not clear how this equipped him to judge cats. Each cat was examined on a card table, which the judges lugged from pen to pen as they awarded the prize money of £57.75, spread among 54 winners. Overall champion was a Persian kitten.

The show was so popular that another was organized later in the year, and shows started to be held in other European countries, where they were also well received by the public.

Twenty-four years later, Madison Square Garden in New York City saw the first-ever American cat show. One hundred and seventy-six cats took part in three categories: Longhair, Foreign Shorthair and Domestic Shorthair, and the prize was won by a Maine Coon. Up until the 1960s, any cat could be shown if three judges agreed that it looked like a registered breed. Now, the lineage of pedigrees is carefully monitored and recorded.

✦ CURIOUSER AND CURIOUSER ... ✦

More like a sausage dog than a cat, the Munchkin is a short-legged breed that most pedigree registries refuse to recognize. Bred from Toulouse, a vertically-challenged tom whose mother lived under a Louisiana trailer, Munchkins are affectionate cats who manage well despite their lack of stature. They certainly have novelty value, but a cat with the build of a dachshund takes getting used to, and the Munchkin will never become a mainstream breed.

Truly a collector's item, the hairless Sphynx looks a bit like Gollum crossed with E.T., with its luminous wide eyes and elfin face, topped off with enormous ears. Rachel in *Friends* had a Sphynx kitten, which led Joey to ask, "How come your cat's inside out?"

The Sphynx's body is covered with a fine peach-fuzz, and it needs regular buffing with a cloth to keep the skin free of any build-up of natural oils that are normally absorbed by the coat. Sphynx-lovers speak fondly of the breed's playful and extrovert temperament, and as it's so rare there's always a long waiting-list for kittens.

WHICH CAT WOULD BUDGE ONLY FOR BUDGE?

Old Mike! Farewell! We all regret you
Although you would not let us pet you,
Of cats, the wisest, oldest, best cat
This is your motto – **Requiescat!**

This delightful and learned elegy was written in 1929 by assistant keeper of the British Museum's Department of Printed Books. In it, he laments the death of Michael, known as Mike, who was carried into the museum one day as a kitten by Black Jack, his predecessor.

Mike attached himself to the museum gatekeeper, and moved into the lodge with him, but would only allow himself to be stroked by two people – his owner, and Sir Ernest A. Wallis Budge, an eminent Egyptologist. Mike's preference was recorded in another verse of his (somewhat lengthy) elegy:

He cared for none – save only two:
For these he purred, for these played,
And let himself be stroked and laid
Aside his antihuman grudge –
His owner – and Sir Ernest Budge!

Budge – author of such illustrious volumes as *The Mummy* and *The Coptic History of Elijah the Tishbite* – wrote a short biographical memoir of Mike, in which he noted that the cat "preferred sole to whiting, and whiting to haddock, and sardines to herring; for cod he had no use whatever."

Mike was buried near the Great Russell Street entrance to the museum, beneath a small tombstone inscribed: "He assisted in keeping the main gate of the British Museum from February 1909 to January 1929."

WHAT DO CATS CATCH?

Most cats aren't that fussy, and will hunt whatever small mammals are available. A study of the stomach contents of domestic cats that were killed on the road (you're not reading this over breakfast, are you?) found that up to 60 percent of pet cats had also eaten prey. The vast majority of these were rural cats, who had feasted on 14 species. The urban cats' stomachs contained mostly pet food – and a single grasshopper.

What domestic cats catch depends on where they live:

Europe: Mice, voles, sparrows, fledglings, shrews, squirrels, spiders and insects
North America: Mice, ground squirrels, flying squirrels, chipmunks, gophers, robins
Sub-Antarctic islands: Terns, penguins and noddies
Australia: Possums, reptiles, ground-nesting birds

What do you get if cross a
cat with a canary?
Shredded tweet.

WHAT ARE THE LAWS OF FELINE PHYSICS?

THE LAW OF INERTIA

A cat at rest will tend to remain at rest, unless acted upon by some outside force – such as the sound of a can opener.

THE LAW OF MAGNETISM

All black coats and navy sweaters attract cat hair in direct proportion to the darkness of the fabric.

THE LAW OF THERMODYNAMICS

Heat flows from a warmer to a cooler body, except in the case of a cat, where all heat flows to the cat.

The Law of Stretching

A cat stretches to a distance proportional to the length of the nap just taken.

The Law of Sleeping

All cats must sleep with people whenever possible, in a position of maximum discomfort for the people, and maximum comfort for the cat.

The Law of Elongation

A cat can elongate her body sufficiently to reach any surface where there might be something to eat.

The Law of Obstruction

A cat will sprawl on the floor so as to be trip-overable from as many directions as possible.

The Law of Rug Configuration

The flat state is unnatural, and no rug will stay in that position for long.

The Law of Resistance

A cat's refusal to comply varies in proportion to a human's desire for her to do something.

✷ HOW CAN WE SAVE THE TIGER? ✷

The flame of the tiger, burning bright, is in danger of being extinguished forever. There are no more than 5,000–6,000 tigers left in the wild worldwide.

Since the late 19th century, 93 percent of tigers' natural habitat has been destroyed. With habitat, goes prey. As the tiger populations dwindle, they are forced to inbreed, which causes health and fertility problems. Not surprisingly, there are 95 percent fewer tigers now than there were 100 years ago.

What can we do to save this glorious creature?

- Make sure we buy coffee from sustainable sources. Illegal plantations eradicate precious tiger habitats.
- Keep a lookout for sellers of traditional Chinese medicine, some of whom are still selling products containing ingredients such as ground tiger bones, which encourages poaching.
- Support tiger conservation projects, which work to protect tigers and save them from extinction.

✷ WHO NEEDS A TAIL? ✷

Not the dear old **Manx**, the cat that caught its tail in the door of Noah's Ark, and has been lacking that feline characteristic ever since! Manx cats originate from the Isle of Man in the Irish Sea and appear on the local stamps and coins. Maybe the first Manx cat was an island cat with a mutating gene – or maybe it swam ashore from the ships of the Spanish Armada. Today the cat is bred as a pedigree and is famous for its depleted hindquarters. There are three levels of Manx tail-lessness:

- Rumpies have no tail at all, and just a dimple at the base of the spine. This is the only variety that can make a career as a show cat.
- Stumpies, Risers or Stubbies have short tails.
- Tailies or Longies have nearly natural, kinked tails.

Despite lacking the usual cat's balancing aid, Manxes are just as agile as other cats and don't seem to have any problem keeping their footing. The breed does have a built-in health hazard, though. If two Rumpies are bred, there's a risk of kittens being born suffering from Manx Syndrome, a deadly condition that affects bowel and bladder.

You'll see ceramic models of the Japanese Bobtail, a traditional good-luck symbol, all over Japan. Known as the *Maneki-neko* or Beckoning Cat, the seated white

cat with a raised left paw predicts great wealth, while a lifted right paw promises happiness and good fortune. The breed has been common in Japan for many centuries, and is seen in ancient prints and paintings.

Their tails are just three to four inches long, and look like furry pom-poms. No two tails are exactly alike, and may be flexible or rigid, curved, angled, kinked or even two-tipped. Forked-tailed cats were demonic, and may have been persecuted in past centuries.

What do you get if you cross a
cat with a tree?
A cat-a-logue!

◦❖ ARE CATS NOCTURNAL? ❖◦

Not exactly. They are "crepuscular," which means they're most active at dawn and dusk. Although cats that have freedom to roam will often pop out for an hour or two in the early part of the night, many domestic pets have adapted to human sleep patterns, and will rest for the entire night, just as their owners do.

❀❖ ARTY CAT FACTS ❖❀

- Cats were used in early Western art as a symbol of treachery and evil, and were traditionally shown sitting at the feet of Judas Iscariot in portrayals of the Last Supper, by artists including Tintoretto and Domenico Ghirlandaio.

- Sinister cat-shaped demons creep around in Hieronymous Bosch's (1450–1516) fantastical paintings.

- As the link with black magic waned, cats began to be portrayed more realistically. "The Skate" by Jean Baptiste Chardin (1699–1779) is almost too realistic. The fish is so slimy you can almost smell it, and the little snarling cat, with its fur on end, does nothing to make the viewer feel comfortable.

- A thoroughly moral series by William Hogarth (1697–1764) published in 1750–51 and called "The Four Stages of Cruelty" shows cats more as victims than evildoers. The series follows the progress of a thoughtless and wantonly cruel young man, Tom Nero, who ends up suffering the same way he has made animals suffer.

- Equine artist second to none, George Stubbs (1724–1806), painted just one study of a cat, "Miss Anne White's Kitten."

- There are dozens of pretty 17th-century painting by artists such as Jean-Honoré Fragonard, Jean-Antoine Watteau and Giovanni Battista Tiepolo, where contented cats snuggle down among their mistress's skirts, or idle their time away in a life of luxury.

- Swiss artist Gottfried Mind (1768-1814) used his cat, Minette, and her kittens as subjects for a whole series of watercolor and pen-and-ink portraits.

- Pablo Picasso (1881–1973) had no truck with painting curled up domestic felines. He said, "I want to make a cat like those that I see crossing the road. They don't have anything in common with house pets. They have bristling fur and run like demons."

- Among David Hockney's (1937-) best known work must be the iconic 1960s painting "Mr. and Mrs. Clarke and Percy," actually painted in 1971, which shows the fashion designer Ozzie Clarke and his wife, fabric designer Celia Birtwell, after their marriage in 1969. Hockney found the painting hard to execute, mainly because he had set out to "achieve ... the presence of two people in this room. All the technical problems were caused because my main aim was to paint the relationship of these two people." Percy, the totem-like white cat perched on Ozzie's lap, wasn't really named Percy. Hockney wanted to use the name for the title, even though it belonged to another of the couple's cats. The cat can be seen as a symbol of infidelity, self-centeredness and freedom to disregard the rules. Not good auspices for a marriage. It didn't last.

❧ WHAT BROUGHT CATS OUT OF THE WILD AND INTO OUR HOMES? ❧

"... the wildest of all the wild animals was the Cat. He walked by himself, and all places were alike to him."

Rudyard Kipling's *Just So Stories for Little Children* sum up the nature of the cat, which lives with humans on its own terms. Cats are never as keen to respond to praise as dogs, because they evolved not as pack animals, but as solitary hunters, for whom food was the only reward worth having. Lone wildcats didn't need to develop social skills, and were content to walk by themselves. Even so, when the opportunity for an easier life presented itself, they were quick to make the most of it.

Domestic cats have snuggled by the hearth for thousands of years. There is evidence of cats being kept as pets from as long ago as 5000 B.C., and certainly by 2000 B.C. cats commonly appeared in Egyptian hieroglyphics and were clearly part of the domestic scene.

Their acceptance of – and by – humans probably began in earnest around 4000 B.C. when the ancient Egyptians first made permanent settlements. Settlers need food, and where food, and especially grain, is stored, it's not long before rats and mice appear. For the indigenous wildcats, all the features of a dream home were there for the taking – warm places to shelter, protection from larger predators who feared humans, and a steady supply of food.

Felis lybica, the African wildcat, wasted no time. With its naturally adventurous and inquisitive nature, it was more than willing to risk taking up residence alongside humans. And so, the first evolutionary step in the rise of the cat was taken.

What makes the difference between a wildcat and a domestic one? Only the more placid, tolerant and unafraid cats would thrive with people, so those characteristics became more predominant. But even so, any cat has the potential to return to a feral state, and the first seven weeks of a kitten's life are crucial.

The feline brain is almost fully matured by seven weeks, and being raised by people affects the development of the young cat's biofeedback mechanisms, which determine how Kitty responds to the world around her. A domestic kitten's brain becomes programmed to stay calm around humans by controlling the flow of hormones that govern the fight-or-flight response. Puss won't panic when she sees or hears people, because she's learned that they're safe. That knowledge, gained young enough, is mentally fixed for the rest of the cat's life.

Over the centuries, wildcats evolved into domestic cats physically, as well as mentally. Their brain size diminished, because a pampered pet simply doesn't need the mental armor a wildcat needs to survive. A more varied diet led to changes in the gut, and the domestic cat has longer intestines than a wildcat. And because house cats don't need camouflage, they flaunt a range of coat colors and patterns that would make them easy pickings in the wild.

But don't be totally fooled. The domestic cat is still wild at heart, and even the softest, most home-loving puss, given half the chance, will follow her nature to stalk, hunt and protect her territory, just as her ancestors did. As Kipling put it:

"The Cat ... will kill mice and he will be kind to Babies when he is in the house, just as long as they do not pull his tail too hard. But when he has done that, and between times, and when the moon gets up and night comes, he is the Cat that walks by himself, and all places are alike to him."

❦ HUMPHREY, THE DOWNING STREET CAT ❦

It's not many strays who end up living at the hub of government, rubbing shoulders, or at least against the legs, of the great, the good and the infamous politicians of the day.

Humphrey, a long-haired black and white cat, turned up on the doorstep of No. 10 Downing Street, London, in 1989. Did Prime Minister Margaret Thatcher personally scoop him up, stroke him and decide to give him a home? We don't know about her feline housing policy, but Humphrey moved into the official post of mouse-catcher, and managed to hang on to his job throughout Thatcher's incumbency (unlike so many of her colleagues).

Next up, John Major, had a soft spot for Humphrey and gave the cat a character reference when he was accused of making lunch out of four baby robins in the garden of No. 10, saying "it is quite certain that Humphrey is not a serial killer." Bill Clinton's two-ton bulletproof Cadillac almost made mincemeat of the cat, who narrowly missed a nasty squishing.

Was our Humph named in honor of Sir Humphrey, civil servant extraordinaire and star of the 1980s political comedy series *Yes Minister?* In that character's words:

"Well, Minister, if you ask me for a straight answer, then I shall say that, as far as we can see, looking at it by and large, taking one thing with another in terms of the average of departments, then in the final analysis it is probably true to say, that at the end of the day,

in general terms, you would probably find that, not to put too fine a point on it, there
probably wasn't very much in it one way or the other as far as one can see, at this stage."

In 1995, Humphrey decided he needed a quieter life and went AWOL. *The Times*
reported his probable death – at which point the staff of the Royal Army Medical
College, a mile away from Downing Street, discovered that their adopted cat, P.C.,
was none other than Humphrey. Humph's return home received international
coverage.

But every cat has his day, and when the Blairs moved in, it was time for Humphrey
to move out. Was he making a political statement? The outgoing Tories thought so.
Or was it just sour grapes when they said, "Humphrey is voting with his paws. After
eight happy years under a Conservative government, he could only take six months
of Labor"? Cherie Blair denied having taken a dislike to him, and Humphrey kept
his thoughts to himself, and spent a happy nine years in retirement living with a
Cabinet Office worker, until he passed away peacefully in 2006.

❧ AS I WAS GOING TO ST. IVES ... ❧

As I was going to St. Ives,
I met a man with seven wives.
Every wife had seven sacks,
Every sack had seven cats,
Every cat had seven kits.
Kits, cats, sacks, wives,
How many were going to St. Ives?

An old riddle – and of course, it's a trick. Only the
speaker was going to St. Ives.

☙ CAN I TEACH MY CAT TO MEOW? ❧

It is possible to teach a cat to respond with a mew when it hears your voice, and it's a useful party trick, too, if you have the kind of cat that likes to place itself in tight, dark places from which it then needs rescuing.

Here's how:

- Start when your cat is young. As with dogs, you can't teach an old cat...
- Use little pieces of your cat's favorite treat. The pros swear by crisply fried liver.
- Find a quiet spot indoors, with no distractions.
- Call your cat's name and give it the food as it comes forward.
- Once the cat has begun to connect hearing its name with receiving the reward, hold the food just out of reach. Many cats will meow at once, especially those that already mew at meal times. Give the food immediately.
- As you continue training, give food intermittently, not every time. Your aim is to phase it out.
- Increase the space between you and the cat bit by bit, until you can call from another room, and the cat will meow. When you can do this reliably, do some training outdoors.
- Never punish or shout at your cat if he doesn't respond.
- Have short, frequent training sessions.

What's the worst kind of cat to have?
A cat-astrophe!

☙ WHAT SNACKS DOES YOUR CAT PREFER? ❧
OWNERS CONFESS ...

- Meg won't eat real food (walks away from a bit of salmon or chicken from my plate). Popcorn, though. She loves popcorn.
- Certified-organic-freerange-additive-free MICE. From under the shed.
- A little diced turkey or yogurt once a day. It's the vet's fault. She made us switch Tabs to a different dried biscuit, so there was hunger strike and I had to bribe her.

- What your cat really wants a is a can of tuna. Oh, go on – just not every day.
- On our boys' menu: spiders, earwigs, moths, butterflies, frogs, and once, a squirrel.
- Black Bum liked broccoli. Preferred it to anything, even tuna fish. I once offered them side-by-side and he went for the veggies. Weird.
- Reese once stole a piece of cheese off my cracker, but his big brother, Apollo, won't even sniff "people food," or cat food. It's dried biscuits or nothing for him. Proves they must taste OK.
- We have a pizza on Friday night, and Whoopy and Mabel get a treat, too. They get their one can of Whiskas of the week, and we split it between them.
- Spike steals salad off unattended plates.
- My ginger cat, Sanji, came up and sniffed my plate last night and started scarfing down my spinach linguine with red pepper sauce. It can't be good.

❧ DO CATS DREAM? ❧

Yes, or at least they show the same kind of brainwaves that sleeping humans produce when they dream. Cats have phases of rapid eye movement (REM) sleep, which is dream-time, and non-REM sleep, when the body is repairing itself.

When the paws start to twitch, the whiskers to whirl and Snoopy is giving little "mrrps," wish him "happy hunting!" and leave him to dream on.

•❂ HOW TO MAKE BOARDING BEARABLE ❂•

Never book your cat into a boarding facility without visiting the place first. The owners may love cats to bits, but still keep them in squalid, dark or cramped surroundings.

- Check that the business is licensed, and ask to see beyond the reception area. A reputable boarder will be happy to show you around the kitty quarters, which should be airy, clean, light and secure. Watch how the owners behave with the pets who are boarding when you visit. Do they greet them by name, stop to stroke them, seem aware of individual animals' little foibles?
- When you've found the cattery of your dreams – or at least one where you can contemplate leaving your cat – book as soon as you've made your own vacation plans. Good facilities fill up early in peak season.
- Make sure your cat's vaccinations for feline enteritis are up to date. Boarders insist that their charges are immunized, and the shots should be given at least six weeks before kitty's vacation.
- When you start packing for your trip, tell your cat she is going on a nice little break as well, and that you will return and bring her home again soon. She may – or may not – understand, but you'll feel better for having owned up before the event.
- Those feline antennae are well-attuned to any change in the usual pattern of life, and will alert their owner to the fact that something's up almost before you've dragged the suitcases down from the attic. The danger is that puss will do a disappearing act at the crucial moment, and you'll have a last-minute panic trying

to find her in time to deliver her to the boarder. On departure day, keep your cat confined in a quiet room, where no one is rushing around trying to find their snorkel.

- Your boarder will provide bowls, bedding and a litter box, but it's a good idea to take some small familiar object along with your cat, like a toy, small piece of bedding or even an item of your clothing.

- You can also leave a supply of your cat's favorite food, or give the boarder exact details. Cats don't take kindly to a change of diet, and may refuse food completely if it's not the right kind. Don't forget to take along medications, vitamins or treats as well.

- Does your cat suffer from car sickness? Remove her food the night before you take her to the boarder to reduce the risk. She won't come to any harm, and having an empty tum might encourage her to nibble on some food as soon as she's settled in.

- Equip yourself with a sturdy basket or pet carrier for the journey. Two cats? Use two carriers. Never attempt to transport a cat in your passenger's arms. Even if she starts off placidly, a sudden noise like a siren could have her flying around the car in a panic.

- When she first goes into her enclosure at the boarder, puss might hiss or snarl at the other cats she can see, or she may take herself off to the darkest corner and refuse to come out. Don't worry. Within 24 hours, most cats have worked out where their new boundaries are, and are happily snuggled into their new bed.

- It's lovely being reunited with your pet at the end of your trip, but don't be surprised if she doesn't fling herself into your arms, purring ecstatically. Cats can be quite
standoffish for a day or two after they've been boarded, and sometimes need time to reacquaint themselves with their homes and their owners. If your cat seems disturbed, you could keep her in for a day or two until she's used to being back home.

"Poor old man. He has fits now, so I call him Fitz-William."

American humorist Josh Billings (1818-1885), remarking to a friend on the state of his dignified cat, William – whose name was never shortened to Bill.

❀❀ WHOSE OWNER WENT UP IN FLAMES? ❀❀

How sinuously a cat can wreathe its way into a story as a symbol of malevolence.
Here are two examples from great writers who added a sinister feline dimension to
their work.

CHARLES DICKENS

In *Bleak House*, Krook, the devilish rag and bone man, owns Lady Jane, a snarling
cat who perches, hissing, on his shoulder and symbolizes her master's brutality.

> *A large gray cat leaped from some neighboring shelf on his shoulder, and startled us all.*
> *"Hi! Show 'em how you scratch. Hi! Tear, my lady!" said her master.*
> *The cat leaped down, and ripped at a bundle of rags with her tigerish claws,*
> *with a sound that it set my teeth on edge to hear.*
> *"She'd do as much for any one I was to set her on," said the old man. "I deal in*
> *cat-skins among other general matters, and hers was offered to me. It's a very*
> *fine skin, as you may see, but I didn't have it stripped off! That warn't like*
> *Chancery practice though, says you!"*

Krook had one of the most gruesomely memorable deaths in fiction and perished as a
result of spontaneous combustion!

EMILE ZOLA

Therese Raquin is a powerful tale of love, lust and murder, to which the tabby cat,
Francois, bears silent witness. At the start of the story he's a simple icon of cozy
homeliness, but soon his watchful presence drives the murderer into a frenzy of guilt
and terror.

Therese is bored with her marriage to Camille, and embarks on an affair with
Laurent, meeting him furtively at her flat, when her husband is out of the way.

> *She would see the cat, sitting motionless and dignified, looking at them. "Look*
> *at Francois," she said to Laurent. "You'd think he understands and is planning*
> *to tell Camille everything tonight. He knows a thing or two about us. Wouldn't*
> *it be funny if one day, in the shop, he just started talking."*
> *This idea was delightful to Therese, but Laurent felt a shudder run through*
> *him as he looked at the cat's big green eyes. Therese's hold on him was not total*
> *and he was scared. He got up and put the cat out of the room.*

Laurent, crazed with desire for Therese, drowns Camille in a faked boating accident. But instead of finding happiness together, the pair are consumed by fear. The sound of the cat's scratching terrifies them both into thinking that Camille's ghost is at the door. In the end, Laurent can stand it no longer.

> *If Laurent had not yet killed the animal, it was because he dared not take hold of him. The cat looked at him with great round eyes that were diabolical in their fixedness. … He was on the point of saying to the cat:*
> *"Heh! Why don't you speak? Tell me what it is you want with me."*

Eventually Laurent hurls the unfortunate François out of the window and kills him. But it's too late. Unable to live with their guilt, the lovers poison themselves.

 Once mounted on a tiger, it is difficult to get off.
Chinese proverb

❧ CAT SPEAK ☙

A 19th-century French researcher Dupont de Nemours spent hours studying the sounds made by dogs and cats, and came to the conclusion that while dogs used only vowel sounds, the cat could enunciate the consonants F, G, H, M, N and V.

It's strange that he doesn't mention the P, R, S and T, the sounds necessary for purring, chirping and hissing. And can cats really say "Fiaow" or "Vrrp"?

Another researcher decided that cats use the same sound or sequence of sounds each time they express the same thought, noting that: "Every emotion has its special note."

Charles Darwin wrote: "Cats use their voices much as a means of expression, and they utter under various emotions and desires, at least six or seven different sounds. The purr of satisfaction which is made during both inspiration and expiration, is one of the most curious. The puma, cheetah, and ocelot likewise purr. But the tiger, when pleased, emits a peculiar short snuffle, accompanied by the closure of the eyelids."

Cats belonging to Madame Michelet, 19th-century cat freak extraordinaire and lady of considerable literary capabilities, didn't just purr, or "*ronron*," as the French would say. She invented a whole new vocabulary for different nuances of feline delight: *mourrons, monrons, mou-ous* and *mrrr*.

·❀ WHOSE BABY ARE YOU? ❀·

Cats are wonderful mothers, who guard their offspring fiercely, educate them in the ways of the world, and nurture them lovingly until they're big enough to fend for themselves.

That maternal instinct is so strong, that a cat who's been deprived of her litter will, on occasion, search around for some other young creature to tend. British naturalist Thomas Brown (1785–1862), recorded this instance of misplaced mothering in his book, *Interesting Anecdotes of the Animal Kingdom.*

> *A cat, belonging to a person in Taunton, having lost her kittens, transferred her affections to two ducklings which were kept in the yard adjoining. She led them out every day to feed, seemed quite pleased to see them eat, returned with them to their usual nest, and evinced as much attachment for them as she could have shown to her lost young ones.*

·❀ HOW TO MOVE HAPPILY ❀·

The arguments for and against an extra bedroom or better ZIP code don't cut any ice with Pugwash. Cats hate change, and will never view the move to a more desirable residence with the same enthusiasm as you do. These tips can help make moving day go smoothly.

- On moving day, confine Pugwash – plus litter box, food and water, bedding and toys – to a room that's been cleared. Put a big notice on the door to warn the movers that this door must stay shut.
- Keep the door firmly shut until the moving van has left and the front door is closed. Take your cat with you to your new house by car.
- At the other end, try to have the contents of one room unloaded first, then put Pugwash – plus litter box, food and water, bedding and toys – in here and leave him to explore quietly.
- When all the furniture's in, the front door is firmly closed and you're reaching for the champagne, let him explore the rest of the house. If he's a timid chap, take him around bit by bit, and let him investigate one room at a time, with you there.
- It's a great idea to keep your cat in for a week or two in a new home – but it's not always practically possible. Do your best, and remember that especially if you've not moved very far, he may well find his way back to his old home. Warn

neighbors to keep an eye out for him, and if he does turn up to let you know, and to harden their hearts and shoo him away. It's the kindest thing.

• Before he has a chance to make a break for it, make sure he is either micro-chipped, or is wearing a collar with the correct address and phone number.

◦⁘ WHERE'D YOU GET THOSE WEIRD EARS? ⁘◦

Shulamith, a stray with strange ears, is the forebear of all American Curls. She was taken in by the Ruga family in California in 1981, and they thought little of her appearance until she gave birth to kittens, two of which shared her curly ears. It's a bizarre breed, whose ears that look as though they've flipped back. The ears curl away from the face and so their owner looks permanently surprised.

In the Scottish Fold, the ears turn over the other way, towards the skull, and in a perfect specimen are set in tight, cap-like folds flat to the head. This makes the face look like a furry owl. The type originated with Susie, a Scottish farm cat, and her kitten, Snooks. Kittens of this rare breed are all born with straight ears, and only a few will go on to develop the characteristic fold when they're about 12 weeks old.

❀❖ WHAT COULD LOOK MORE APPEALING? ❖❀

Cats are just as popular as decorative devices today as they ever were. Through the ages, they've been used to enhance a fantastic range of items, from precious artifacts to oven mitts.

- Egyptian jewelry – cats adorn ancient bracelets, rings, amulets, cosmetic jars.
- Cat figures, ranging from stone statues of lions, to simple earthenware statuettes of domestic cats, have always been well-liked. Cat-shaped ornaments for the home became fashionable from the 17th century onward. Made to sit on a shelf or mantelpiece, many were cheap, everyday pieces, but museums have Faberge cats, German Meissen cats, Dutch Delftware and Ch'ing cats from China. Other examples include a cat-shaped bottle from ancient Persia, Italian vases painted with cat motifs, and feline Toby jugs.
- Cats and lions have been heraldic emblems for hundreds of years, and appear on many coats of arms and family crests.
- Walter Potter, a British taxidermist, created a macabre display with his "Kittens' Wedding" – a whole congregation of stuffed kittens, dressed up in little frocks and suits. The Victorians loved it apparently, but to the 21st-century eye it's downright creepy.
- Cats make wonderful toys. From pocket-sized mini-pusses to almost life-sized stuffed leopards, there are cuddly cats to suit every age. Steiff, famous German makers of teddy bears, also make upmarket toy cats in colors from tabby to purple.
- Victorian children enjoyed playing with metal nodding cats. Colorful wind-up tin cats are still made in Eastern Europe and China.

CATS THAT SERVED ON SHIPS IN THE NAVY		
Annie, HMS Anson	Frankenstein, HMS Belfast	Stripey, HMS Warspite
Beauty, HMS Black Prince	Hoskyn, HMS Chester	Thomas Oscar, HMS
Blackie, HMS Prince of Wales	Jimmy, HMS Renown	Scorpion
(renamed Churchill after	Minnie, HMS Argonaut	Tommy and Lucky,
Winston noticed him when the	Peggy, SS Julie	Empire Winnie
Prime Minister came on board	Side Boy, HMS Neptune	Whisky, HMS Duke
for a meeting with President	Scouse, HMS Exeter	of York
Franklin D. Roosevelt)	Smokey, HMS Bulldog	

- Cat postcards are fun to collect. Look for caricatures by Arthur Theile and Louis Wain's anthropomorphic cats, all dressed up and behaving just like people do. Fashions move on, in postcards as in dress. In the 1930s, cards with shots of different breeds were popular, while in the 1950s, little girls cuddling kitties caught the public fancy.
- Cat stamps are enormously popular, and have been produced by numerous countries, from Albania to Slovenia, through the Philippines, Austria, Australia and Africa.

Great A, little a, bouncing B,
The cat's in the cupboard, and she can't see.
A traditional rhyme, teaching
the letters of the alphabet.

❖ WHAT'S IT LIKE TO BE A CAT? ❖

It's fun being Biscuit! Just imagine you could change places with your cat. While she's figuring out the can opener, this is what your world will be like.

- You can spring to a height of 10 feet without even thinking about it. Hmm, the top of the fridge could do with a dusting.
- You go into a room and instantly your nose tells you who's in there.
- The world is a place of dull colors and blurred edges, but there are compensations.
- One twitch of your ears picks up the tiniest high-pitched sound, and pinpoints it precisely.
- Your furry face is surrounded by a force-field of finely tuned whiskers, which detect the lightest breeze.
- At night, you see as if you had a built-in infrared camera. No more stumbling around in the dark.
- Your body is insulated from hot and cold, but your lips and nose are as sensitive as a human's fingertips.
- You're so supple you can get your leg over your head, and you can twist around to touch your back with your tongue. If your yoga teacher could see you now!
- If you fell off the top diving board you wouldn't belly-flop, but you'd flip over in mid-air and splash-land feet first.

❧ WHY DO CATS GO CRAZY IN THE EVENING? ❧

Up the stairs, down the stairs, across the sofa-back – mind your head! – three times around the rug, wheeee, over the table, is that my tail – gotcha! – and off up the stairs again …

Almost all cats occasionally indulge in the feline evening-mad-dash of crazy pouncing, trouncing and whirling around. Kittens really love to go totally crazy, but older cats can't always be bothered to rouse themselves, and once Alphonse passes his fifth birthday he'll only perform on very special occasions.

What makes a cat give in to an attack of the whirling dervishes? Like so much about feline behavior, we can only guess, but it's probably a way of unleashing the energy domestic cats don't need to use to survive, coupled with a bit of hunting practice. Cats generally sleep during the day, but they're programmed to come alive in the evening, and that's when they're most likely to bounce off the walls for a bit, before sitting down for a restorative wash.

❧ HOW DID THE CAT'S EYE IMPROVE ROAD SAFETY? ❧

Percy Shaw (1890–1976) became a multimillionaire, and saved thousands of lives to boot, with his invention of the cat's-eye reflective road-marker.

Although Shaw himself sometimes dismissed the story, it's popularly believed that he came up with his lucrative idea one night in 1933. In those days, drivers relied on the reflection from their headlights on the trolley tracks set into the road to keep them on course, but the tracks had been taken up for repair. Shaw was driving home through fog so thick that it masked the edge of the road – and the steep hillside drop just beyond it.

Just as he was getting seriously worried, Shaw saw his headlights reflected back in the eyes of a cat that was sitting by the road, a glimmer of light that sparked an amazing idea – to reproduce the same effect in a form that could be installed on any dark and dangerous stretch of road.

Shaw spent several years perfecting his product, a robust, permanently fixed device needing minimal maintenance and self-cleaned by rainwater. In 1937, he won a competition staged by the Department of Transport, who were concerned by the rising number of night-time crashes as more and more people learned to drive. But business didn't really take off until World War II, when the dangers of driving in the air-raid blackout made it imperative to install many more cat's eyes.

Shaw's factory continued to thrive after the war, and his fortune was made. He was a mildly eccentric figure, whose house was bare of curtains, carpets and other comforts, but boasted two Rolls-Royces parked outside and a cellar stacked with crates of Shaw's favorite beer, Worthington's White Shield. Thanks to him, to the extraordinary reflective properties of the feline eye – and to that one roadside cat – thousands of night-drivers every year are still saved from death and injury by his invention.

✺ CAN I WALK MY CAT ON A LEAD? ✺

You know that Sparkle would love to go out, but as an indoor cat, there's simply nowhere she can go for a breath of fresh air. No guarantees, but you could give this a try:

- Use a cat harness rather than a collar, which might hurt Sparkle's neck or even allow her to slip out.
- Start while she's a kitten. Older cats are much more likely to resist being strapped in.
- First put on the harness minus the lead, and let her get used to that. Distract her with play.
- Keep the harness on for a short time to start with, and give Sparkle a reward when you take it off.
- Now attach the lead, and simply leave it to trail. Never leave your cat alone while she's wearing either harness or lead.
- Take the end of the lead, and let her get used to you holding it and following her.
- When she accepts that, and is willing to follow you, try it outdoors. Choose a quiet spot, and increase the time gradually.
- Never walk near busy roads, or in parks where dogs might roam. Sparkle's instinct is to take flight if she's frightened, and she could get lost or hurt.

❧ THE CAT'S PAJAMAS ... ❧

Who do you know who's the cat's pajamas – or, to put it another way, who's great, wonderful, stylish, the best? This 1920s saying, from the age of flappers and Broadway nightclub life, successfully crossed the Atlantic, and is as well known in the U.K. as it is in the States.

Puss doesn't wear PJs? *The American Dictionary of Slang* lists other feline expressions: "the cat's eyebrow, ankle, adenoids, tonsils, galoshes, cufflinks and roller skates" – that you can use to describe something that's simply beyond the bee's knees.

❧ WHY DO CATS SLEEP SO MUCH? ❧

Not for Aladdin the torments of insomnia. He can lapse into a delicious slumber seemingly at will, and in the most unlikely spots. And so he does, for up to 16 hours a day, longer if he's a kitten or senior cat.

A cat's internal clock is governed by the hunting impulse, and even the best-fed domestic beast will probably be on full alert at dawn and ready to prowl at dusk. This is why it can be a mixed blessing to let Aladdin sleep on your bed, unless you like having your head pummelled at 5 a.m. If his early waking becomes a problem, try installing heavier curtains, or even black-out blinds in your bedroom, so the first rays of sun don't wake him.

His sleepiest times are:

- The middle of the day, when prey and predators alike snooze through the heights of the sun until it's cool enough to be active again.
- The depths of the night, after the hurly-burly of the late evening hunt.
- After a good meal, when cats, just like their owners, feel pleasantly heavy-lidded.

Cats seem to be even sleepier in hot weather, and will also pass the time horizontally if there's no one around who wants to play or chat with them. A cat that appears to be deeply unconscious will quickly revive when the family comes home and there's a chance for a bit of fun.

Aladdin also likes to amuse himself by lapsing into standby mode, with just a slit of an eye open so as not to miss anything. At a glance he's asleep, but look closer, or stop to caress him, and he'll rouse himself enough to enjoy the attention before slumping back into an entrancing state of semi-somnolence.

Other reasons why cats need to catch up on their shut-eye? Maybe it's a time for dreaming about how to cover more of your belongings with fur, or they simply need the rest because they were busy ruling the world, while their owners slept.

❧ FELINE LIGHT METER ❧

In the early days of photography, light was essential. You could only take a decent picture on a sunny day, but it was hard to decide just how much light was enough. Nor was there such a thing as a quick snap. Setting up a shoot took time.

Oscar Rejlander (1817–1875) hit on the idea of using a cat as a primitive light meter. He persuaded puss to sit in the spot that the model would occupy (history doesn't record how he made the cat stay put), then looked at its eyes to judge by how much reflection they caught from the sun whether there was any point in embarking on the picture.

☙ DO SIAMESE CATS COME FROM SIAM? ❧

Perhaps this best-known breed did spring up in Siam, now Thailand, but it's just as likely that it originated in China. The Siamese stayed in the East for centuries, and didn't make an appearance in the West until Victorian times, when the king of Siam sent a pair, Pho and Mia, to the British consul general.

In 1885, Pho and Mia's kittens were shown at the Crystal Palace Cat Show in London, where they won prizes. Today they'd be lucky to come in last among the also-rans, against the sleek and elegant Siamese puss that pedigree-lovers appreciate today. That fine-boned and sophisticated shape is the result of systematic breeding, using only the very finest cats. The Victorian kitties were more thickset, with rounded heads, kinked tails and a pronounced squint. Interesting, but not the like the svelte 21st-century felines.

These first Siamese immigrants were seal-points, with creamy coats and brown face, tail and paws. Blue-points, whose markings are gray and whose coats look slightly blue, followed in 1890, and chocolate-points, whose markings are lighter, made the trip in 1900. Now you can virtually pick point colors from a color chart – lilac, cream, tabby, red – as interbreeding has produced so many different variations.

☙ FROM POOR MATTHIAS ☙

Cruel, but composed and bland,
Dumb, inscrutable and grand,
So Tiberius might have sat,
Had Tiberius been a cat.

Matthew Arnold (1822–1888)

☙ CAN CATS PREDICT EARTHQUAKES? ☙

Tales of weird cat behavior in the run-up to an earthquake have been around for centuries. In 373 B.C., history relates that all the animals, including rats, snakes and weasels, fled the Greek city of Helice just days before it was devastated by an earthquake.

Many pet owners in quake zones have reported that their cats became anxious and restless, or hid themselves away before an earthquake arrived. But is it true?

Do animals detect tiny warning tremors, too slight to be noticed by humans? Can they pick up electrical changes in the atmosphere? No one knows.

Earthquakes' unpredictability makes them supremely deadly. There are about half a million quakes every year, of which 100 are destructive. Countless deaths and enormous damage could be prevented if scientists could pin down whether animals detect impending quakes, and if so, how they do it.

At present, there is nothing stronger than anecdotal evidence to go on. There have been a few studies, but no concrete results. It's virtually impossible to perform a controlled experiment on animals, because they react to all sorts of things, all the time.

Some skeptical scientists are inclined to put the whole thing down to the "psychological focusing effect," which happens when people look back with hindsight, following a cataclysmic event. From that perspective, they recall behavior as being "odd" that wouldn't have struck them as abnormal if nothing had happened after it.

One way to carry out research would be to set up a Web site where people in earthquake zones could log strange pet behavior. A surge of e-mails might show if an earthquake was about to happen. The information could be computer-checked against other factors that affect animals, such as high winds, and used in conjunction with seismological measuring devices.

But it needs someone with the vision, and the funds, to find out once and for all, if cats really can predict quakes.

❧ CAT RITUALS AND BELIEFS ❧

- Ancient Egyptians buried cat-shaped amulets with their dead to ward off evil spirits, and thought a vast celestial cat inhabited the heavens. Its eye was the moon. Their cat cult attracted huge followings to Bast, goddess of fertility and motherhood. Numerous sacred cats lived in her temple at Bubastis and thousands of Egyptians made their way there to the annual festival, where there was feasting and cats were sacrificed. Dead cats were lamented over, and taken to a sacred house to be embalmed. Even a humble worker's cat would be carefully prepared for burial, and sent on its way with pots of milk and even mummified mice. The cat cult ended in A.D. 390, when an imperial decree outlawed Bast.

- South Americans did not worship the domestic cat, but believed that jaguars were the spirits of dead medicine men.

- In Mexican Zuni mythology, the cougar, because of its power, strength and steadfastness, was the message-bearer between humans and the higher spirits, Mother Earth, Father Sky and the Originator of All.

- Nordic fertility goddess Freya rode through the air on a chariot drawn by a team of giant cats, which symbolized her two attributes of fecundity and ferocity.

- Finns believed that, when they died, their souls would be borne away on a cat-drawn sledge.

- Ancient Persian legend tells how the cat was born – when the lion sneezed.

- The Islamic prophet, Mohammed, owned a cat, Muezza, whom he treated so tenderly that, when she fell asleep on the sleeve of his robe, he cut away the piece where she lay rather than disturb her. It's also said that a cat once saved Mohammed from being bitten by a snake. Cats are still allowed to wander in an out of mosques at will.

- The Parsees, an eighth century religious group, revered the cat and considered it a crime to kill one.

- In Buddhism the cat, although now permitted to strive towards nirvana like all other living creatures, was originally left off the list of protected animals because it was the only animal, apart from the snake, not to weep when Buddha died, and it had the nerve to eat the rat that had gathered in mourning with the other beasts. Cats are kept to control mice in Buddhist temples across Asia. Many have the typical pointed coat pattern of the Siamese, which is said to have come down from these temple cats.

- Chinese Buddhists believed the cat could fend off evil spirits that only its all-seeing eyes could discern in the darkness. Statuettes of cats were displayed in homes throughout China to keep rats away.
- When a cat died in Japan it was buried near the Buddhist temple, and its owner made an offering of a painting or sculpture of the late lamented pet to ensure a lifetime's good luck.

☙ GOODNESS GRACIOUS, WHAT IS THAT? ❧

Outspoken Conservative MP and former Shadow Home Secretary Ann Widdecombe has always loved cats, and speaks fondly of her childhood pets – Tibby, Jimmy and Mitten-the-Kitten.

As an MP she had two "Westminster cats," Sooty and Sweep. Sweep lived to be 24, long enough to become a "ministerial cat" when his owner was promoted from the back benches of the Houses of Parliament.

Ann was too busy to have cats for some years after this, but when her mother moved in, they chose two companion cats for her from a cat sanctuary – Carruthers, an old cat, and Pugwash. Unfortunately, neither of these two lived very long, and after they died, Ann wrote this brief but moving poem in their memory:

> *Goodness gracious what is that*
> *It's Mr. Pugwash, my black cat*
> *Good gracious are there others?*
> *Yes indeed, my cat Carruthers!*

Perhaps fiction is a greater strength than poetry. Ann has since published four novels, and replaced the late lamented cats with two more, Pugwash II and Arbuthnot.

> "
> *All cats love fish but fear to wet their paws.*
> An old saying referring to a person who wants to
> get their hands on something valuable, but doesn't want to
> take risks or put themselves to any bother.
> Shakespeare refers to the saying in *Macbeth*:
> *"Letting 'I dare not' wait upon 'I would,'*
> *Like the poor cat i'the adage."*
> "

❧ WHY DID THE BRIDE JUMP OUT OF HER SKIN? ❧

As she arrived outside the Cathedral Parish Church of St. John's, Brecon, and paused to arrange her train, a cat, Buonaventura – Boni for short – plummeted 10 feet from the tree where he'd been quietly watching and engaged in battle with her bridal train. The photographer had to extricate Boni from the gown, so that the bride, only slightly disheveled, could make her entrance as planned.

> *I saw a Puritan-one*
> *Hanging of his cat on Monday,*
> *For killing of a mouse on Sunday.*
>
> From *Barnabee's Journal Part 1* by Richard Braithwaite
> (1588–1673)

❧ HOW DOES ONE BREED TURN INTO TWO? ❧

When it's a Burmese, the breed that originated from the American mating of a cat from Burma (which turned out to be a Tonkinese, but let's not complicate things even more) and a Siamese.

American Burmese have a rounded face with chubby little cheeks, golden marbles for eyes, a friendly and relaxed temperament and are generally so cute you could eat one for lunch.

Their European cousin – if cousin is the right word – looks quite different. The European breed is descended from the American line, but European breeders wanted a more Oriental look and a bigger selection of coat colors. Deliberate and accidental crossbreedings with, among others, a ginger tabby, a tortoiseshell-and-white farm cat, and a red-point Siamese produced a fantastic hodgepodge of shades, including Brown Tortie (dark brown speckled through dramatically with red and cream), Chocolate Tortie, creamy Chocolate, Lilac, Brown, Cinnamon, Fawn, Blue and Red (more of an orangey-cream and much-prized). Delicious!

The cat itself looks quite different from the American version. The head is blunt and wedge-shaped, the ears relatively large and with a rounded point, the body strong, well-muscled and sturdy, eyes set at a slight slant and golden yellow or green. Again, it's a friendly, sociable cat, with a loving and playful nature, the perfect pet for a family home, where there's a lot going on and always someone around to offer a cuddle or lap.

❧ WHAT DO CATS DO ON YOUTUBE? ☙

Bored? Depressed? Here's how to while away half an hour alternately hooting with laughter and saying, "aaah, how sweet," as you watch online video clips of cats doing everything you can imagine, and more …

Go to www.youtube.com, search for "funny cats" and prepare to laugh until you cry at cats cannoning into the camera (yowch!), extruding themselves from the innards of a sofa, beating a cactus to death, sliding off counters, walking on their paws, leaping into a bathful of water and out again.

Or just hunt through the most recent uploads, and see the latest that cat owners have captured on video. The offerings change constantly, but this selection gives you an idea of what to expect:

- Nora, the piano-playing cat, who strokes the keys lovingly and appears captivated by the "music" she makes.
- A version of Hamlet with cat heads superimposed on human bodies. A trifle spooky.
- Kittens dancing.
- Cats flushing the toilet, using door handles, switching lights on … and off … and on again.
- Scare-haired cats putting the fear of the devil into themselves, with their own reflection.
- Cats who "talk" to their owners.
- Kitty falling asleep – very sweet, and gives "dropping off" a whole new meaning.
- Cats boxing and biffing each other.
- Feline weight-lifting (with fish).

❧ FROM JUBILATE AGNO ❧

For I will consider my Cat Jeoffry.
For he is the servant of the Living God, duly and daily
serving Him.
For at the first glance of the glory of God in the East he
worships in his way.
For this is done by wreathing his body seven times round
with elegant quickness.

For he counteracts the powers of darkness
by his electrical skin and glaring eyes.
For he counteracts the Devil, who is death, by brisking
about the life.

Both from *"Jubilate Agno"* by Christopher Smart
(1722–1771)

❧ WHO WENT FROM CATWALK TO CAT SANCTUARY? ❧

Former top model Celia Hammond admits to being obsessed with cats. "You just become cat mad. … They don't give their heart to just anybody, like dogs. … Is there anything so wrong about getting emotional feedback from an animal?"

At the height of her fame in the 1960s, Celia's svelte beauty featured on the cover of *Vogue*, and she traveled all over the world on modeling jobs. After seeing film of the Canadian seal hunt, she stopped modeling fur, and began to campaign against the fur trade.

About that time, she also became aware of the plight of feral cats, and began helping to capture and neuter them, before returning them to their own environment. At that time, stray cats were regarded as being little more than vermin, and Celia began – and continues – to campaign for humane treatment for strays.

She set up CHAT, the Celia Hammond Animal Trust, and operates her own cat sanctuary, where cats rub shoulders with once mistreated chickens, pigs, ponies and other animals. Many cats are re-homed, but those that can't easily find a new owner because of their temperament, age or appearance, stay on at the sanctuary, where they're allowed to roam freely on the 12 acres of grounds, and sleep in snug chalets.

It's all a far cry from the world of high fashion, but Celia is devoted to helping animals, and in 2004 received the RSPCA's most prestigious award for her dedication to animal welfare.

❧ WHO ARE THE FATTEST CATS? ❧

- *"And to my faithful cat, a lifelong pension ..."*
  Both the 18th-century Earl of Chesterfield and the Second Duke of Montagu left their pets suitably well-heeled.
- Charlie Chan – a cat, not a detective – inherited a house and its contents worth a cool quarter of a million dollars.
- One American oil heiress left a fortune to found a home for 125 New York strays.
- The wealthy Mrs. Walker left a cat charity $5 million, as long as they looked after her pet for life.
- Ben Rea left $12 million to cat charities – and his pet cats.
- The richest cat ever was Blackie, who collected £15 million when his owner departed this life.

❧ CAN YOU STOP CATS HUNTING? ❧

Can you stop the sun shining? Hunting is second nature to a cat, although some are keener stalkers than others. But if your cat is a menace to the local bird population, there's plenty you, and your neighbors, can do to minimize the carnage.

CAT OWNERS

- Don't let your cat out in the early morning or evening, when birds are at their most active and come down into gardens to feed.
- Make sure your cat has a collar with a bell.
- Keep puss well fed and cared for. This in itself won't stop her hunting, but should encourage her to stay close to home.
- Fence your garden. Cats can get over fences, of course, but do everything you can to make it harder. You could even consider installing a solar-powered electric wire along the top.
- Neuter your cats and you'll discourage them from straying, as well as doing your bit to control the cat population.

PROTECT WILDLIFE

- Neighbors could fit an ultra-sonic deterrent (available from the Royal Society for the Protection of Birds) which emits a sound that humans can't hear and can detect movement to a range of about 40 feet.
- Position prickly plants such as holly or pyracantha beneath the bird table. Or plant a few strategic "Scaredy Cat" plants, the *Coleus canina*, which smells foul to cats (and dogs and foxes), but not to humans (unless they touch it). It's an attractive foliage plant, with blue flower spikes in summer.
- Put feeders, bird boxes and bird tables up high and position them away from surfaces that cats could use as a jumping-off point.
- Fit nest boxes that have pointed roofs, not flat, so cats can't sit on them and wait for the unsuspecting parent birds to emerge.

There are two means of refuge from the miseries of life: music and cats.
Albert Schweitzer (1875–1965), missionary surgeon and Nobel Peace Prize winner

✿❖ WHAT ARE THE RISKS OF
AN OPEN DRAWER? ❖✿

Poet William Cowper knew them well, as this extract from "The Retired Cat"
(1791) makes clear. The poem is about the poet's cat, who was, almost fatally, "much
addicted to inquire/For nooks to which she might retire …"

… But love of change, it seems, has place
Not only in our wiser race;
Cats also feel, as well as we,
That passion's force, and so did she.
Her climbing, she began to find,
Expos'd her too much to the wind,
And the old utensil of tin
Was cold and comfortless within:
She therefore wish'd instead of those
Some place of more serene repose,
Where neither cold might come, nor air
Too rudely wanton with her hair,
And sought it in the likeliest mode
Within her master's snug abode.

A drawer, it chanc'd, at bottom lin'd
With linen of the softest kind,
With such as merchants introduce
From India, for the ladies' use –
A drawer impending o'er the rest,
Half-open in the topmost chest,
Of depth enough, and none to spare,
Invited her to slumber there;
Puss with delight beyond expression
Survey'd the scene, and took possession.
Recumbent at her ease ere long,
And lull'd by her own humdrum song,
She left the cares of life behind,
And slept as she would sleep her last,
When in came, housewifely inclin'd

The chambermaid, and shut it fast;
By no malignity impell'd,
But all unconscious whom it held.

Awaken'd by the shock, cried puss,
"Was ever cat attended thus!
The open drawer was left, I see,
Merely to prove a nest for me.
For soon as I was well compos'd,
Then came the maid, and it was clos'd.
How smooth these kerchiefs, and how sweet!
Oh, what a delicate retreat!
I will resign myself to rest
Till Sol, declining in the west,
Shall call to supper, when, no doubt,
Susan will come and let me out."

The evening came, the sun descended,
And puss remain'd still unattended.
The night roll'd tardily away
(With her indeed 'twas never day),
The sprightly morn her course renew'd,
The evening gray again ensued,
And puss came into mind no more
Than if entomb'd the day before.

With hunger pinch'd, and pinch'd for room,
She now presag'd approaching doom,
Nor slept a single wink, or purr'd,
Conscious of jeopardy incurr'd.

That night, by chance, the poet watching
Heard an inexplicable scratching;
His noble heart went pit-a-pat
And to himself he said, "What's that?"
He drew the curtain at his side,
And forth he peep'd, but nothing spied;
Yet, by his ear directed, guess'd
Something imprison'd in the chest,
And, doubtful what, with prudent care
Resolv'd it should continue there.
At length a voice which well he knew,
A long and melancholy mew,
Saluting his poetic ears,
Consol'd him, and dispell'd his fears:
He left his bed, he trod the floor,
He 'gan in haste the drawers explore,
The lowest first, and without stop
The rest in order to the top;
For 'tis a truth well known to most,
That whatsoever thing is lost,
We seek it, ere it come to light,
In ev'ry cranny but the right.
Forth skipp'd the cat, not now replete
As erst with airy self-conceit,
Nor in her own fond apprehension
A theme for all the world's attention,
But modest, sober, cured of all
Her notions hyperbolical,
And wishing for a place of rest
Anything rather than a chest.

❧ WHO'S THE MIGHTY MOUSER? ☙

A computer engineer is summoned to a woman's house. She's switched from a wireless mouse to using the tracker pad on her laptop, but every time she uses the computer the cursor goes wild, careening around the screen and clicking at random. Either the electronics have gone haywire, or she's got a poltergeist.

The engineer inspects various wires – no problem. He scratches his head, and, from the corner of his eye, glimpses a little red light, coming from behind a pile of books on the floor.

Engineer moves books, and behind them is a small cat, batting an optical mouse around and smacking the buttons. … "Oh, there you are Sammy," says the laptop lady. And tells the computer man that Sammy used to warm himself on her monitor, but since she bought a flat screen he's been a bit disgruntled.

Seems Sammy has secretly swiped the old mouse off the desk and hidden it behind the books, where he can torment his owner whenever she switches on the computer. Not deliberate, of course.

CATS AND COMPUTERS 6
Why does my human get so irate
when I sdklfj iuf ap adkfj'-0=

❧ WHICH PLANTS ARE POISONOUS TO CATS? ☙

Beware the vase of lilies, the pot of azaleas, the ivy-clad wall. All of these plants can do your cat serious damage if she eats them. Having said that, most cats like grass the best, and will turn their noses up at anything else growing in the garden, although a flower that's been carefully arranged in a vase could prove more enticing.

Here are just a few of the most common plants toxic to cats.

- Azaleas are moderately toxic.
- Chrysanthemum stems and leaves can cause dermatitis, but most cats are put off by the acrid scent before they get that far.
- All parts of the daffodil or narcissus can cause vomiting or diarrhea.
- Hydrangea flowers could give Achilles a tummy ache.
- Iris bulbs are potentially harmful, but the flowers and foliage are not dangerous.
- Ivy is a definite no-no, and could even prove fatal if your cat ate a lot of leaves.

- Marigold leaves and stems can cause an upset tum, but again the pungent smell acts as a useful deterrent.
- Any part of the lily is highly poisonous to cats, and could even cause kidney failure and death. Don't buy the cut flowers, or grow them.
- Wisteria seed pods drop to the ground like rain when the flowers are wilting, but be sure to sweep them up before Achilles goes out for a stroll or, better still, deadhead this trailing plant thoroughly so the poisonous pods never have a chance to form.

> *If animals could speak the dog would be a blundering outspoken fellow, but the cat would have the rare grace of never saying a word too much.*
>
> Mark Twain, American author of
> *The Adventures of Huckleberry Finn*

❧ ARE BEETLES GOOD FOR CATS? ❧

Insects are fair game to a born hunter, even though they don't taste that good, and you're quite likely to catch your cat crunching on an earwig, or see it leap up on a summer's evening to capture a moth on the wing – and swallow it at a gulp.

Nineteenth-century naturalist Philip M. Rule, wrote this in his book *The Cat: Its Natural History, Domestic Variations, Management and Treatment*:

> *It may here be observed that the cat is even sometimes of a slightly insectivorous propensity. Young, sportive cats, more especially, have much amusement in playing with cockroaches, and sometimes eat them. But they appear to eat them more from accident or idleness than from desire.*
>
> *Occasionally, pussy will be fortunate in catching such rare game as a cricket. Flies are not easily caught, except in a window; and they are said to make cats thin. Beetles, I think, do a cat no harm.*

MAY I HAVE ONE WITH MATCHING PAWS, PLEASE?

Cat breed associations stipulate the correct color for what they call the "leather" on nose, lips and paw pads:

White coat/pink leather
Black coat/black leather
Blue (gray) coat/blue leather
Ginger coat/pink to brick-red leather

MARIGOLD

She moved through the garden in glory, because
She had very long claws at the end of her paws.
Her back was arched, her tail was high,
A green fire glared in her vivid eye;
And all the Toms, though never so bold,
Quailed at the martial Marigold.

This short word picture of Marigold, by Richard Garnett (1835–1906), conjures up a wonderful vision of a queenly cat.

WHY DO PLUMP CATS HAVE PROBLEMS?

A weight problem is just as harmful for a cat as it is for a person. Most regulate their own food intake and only eat when they're hungry, but the food-loving or over-indulged few can easily balloon. Overweight cats end up with similar health problems to humans and can develop diabetes, heart disease and other nasty illnesses which give them a shorter life expectancy.

Guinness World Records stopped documenting the fat feline record, when they realized overly ambitious – some would say cruel – owners were overfeeding their pets in a bid to win.

Eighteenth-century Scottish artist John Kay painted his enormous pet, who at the time he was the largest cat in Scotland, overflowing a chairback.

Himmy, a portly Australian puss, looked as if he was all belly and no legs. At 46 pounds, he was so obese that his owner had to trundle him around in a wheelbarrow. Predictably, he died of heart failure.

HOW DOES AN ARTIST AVOID BEING TWEE?

Cats are an abiding inspiration for Scottish artist Elizabeth Blackadder (1931–) the first woman to be a member of both the Royal Academy of Art and the Royal Scottish Academy.

Best known for her distinctive still-life paintings of flowers and decorative objects, cats began to make a casual appearance in Elizabeth's work from the late 1960s. Initially, she found it hard to capture their quintessentially feline qualities without lapsing into sentimentality. "I looked at other people who'd painted cats beautifully … like Bonnard, Gwen John and Manet. … And I thought, it is possible."

Cats stroll in and out of her delightful watercolor still lifes and portraits. Sometimes it's the face of her Abyssinian cat that peeps over the edge of the frame into the picture to gaze unblinkingly at the viewer. Or the back of the tortoiseshell's head just shows in the corner of a canvas, ears twirling interestedly as the cat looks into the composition. At other times, only the hindquarters and tail appear, as their owner takes off, out of the picture.

As Elizabeth's work has progressed, cats have come to play a larger part in her work and the artist uses their color and form to offset other shades and shapes. A black cat contrasts with the brilliance of pink orchids, or the sleek form of the Abyssinian appears like a jungle cat glimpsed through undergrowth, as its body is partially masked by the drooping fronds of iris leaves.

ARTISTS WHO USED CATS IN THEIR WORK

Pierre Bonnard (1867–1947)

French painter known for his intense, mosaic-like use of color in richly textured impressionistic pictures of sunlit rooms and landscapes. His wife of 49 years, Marthe, appears over and over again in his work, occupied in her everyday, domestic concerns "washing or bathing, talking to a dog or a cat, sitting dreaming over a cup of coffee."

Alberto Giacometti (1901–1966)

The Swiss sculptor, known for his characteristic ultra-thin sculptures of figures, was asked which piece of his work he would save if his house burned down. He replied, "If there was a cat, and my works, I would save the cat. A cat's life is more important than art."

Paul Klee (1879–1940)

Klee was a Swiss artist known for his exquisite use of color in his delicate kaleidoscopic chequerboard paintings. He loved cats, and occasionally painted them in rather alarming close-up, almost as if the viewer is being sized up as prey. A book by Marina Alberghini, *The Cosmic Cats of Paul Klee*, included the artist's photographs of some of his cats, including the long-haired Mys, tabby Fritzi, a kitten named Nuggi and the white, long-haired Bimbo.

Pierre Auguste Renoir (1841–1919)

Renoir often included cats in his paintings, showing them being playful, or in semi-erotic poses with his female models in paintings such as "Girl with Cat."

Jean-Claude Suarès (1942–)

Suarès, New York cartoonist, author, photographer and illustrator, and prolific author of cat books including *Fat Cats*, *Funny Cats* and *Cats In Love*, said of his Egyptian upbringing: "I was never allowed a cat or a car in my homeland, so when I came to the United States I got a cat and a Rolls-Royce. Then I started to multiply everything. … My record number of cats is 13 and my record number of Rolls-Royces is 19."

SANDY SKOGLUND (1946–)

In 1980, surrealist photographer and artist Sandy Skoglund created an installation named *Radioactive Cats*, which featured about 30 lime-green life-sized models of cats, made from chicken wire and plaster, seething around a gray kitchen scene, complete with fridge.

ANDY WARHOL (1928–1987)

The American avant-garde painter and photographer Andy Warhol, and his mother, Julia Warhola, adored cats and owned so many that they often resorted to giving kittens away to their friends. Warhol extended his keenness for repetition, seen in his famous works of Campbell's soup cans and Marilyn Monroe, to the naming of cats. His were all called Sam except for one – Hester.

In the 1950s, Warhol created two privately printed books of cat drawings, most of which were given to friends and clients. They were called *Holy Cats by Andy Warhol's Mother* and *25 Cats Name Sam and One Blue Pussy*. Warhol's mother produced the calligraphy in the second book, and accidentally left the "d" off the word "Named" in the title. Warhol thought random mistakes could enhance his work, and so decided to keep the typo. It's not the only error. The book actually contains only 16 cats named Sam.

☙ WHICH CAT IS A GORILLA'S PET? ❧

Koko, a female gorilla born in 1971 and kept in captivity in California, was taught by Dr. Francine Patterson to communicate in sign language.

Koko apparently asked her trainer if she could have a pet, and when introduced to a litter of abandoned kittens, picked out a gray male Manx, and named it All Ball.

She looked after the kitten as if it were her own baby, but only a few months later, All Ball escaped from Koko's cage, and was knocked down and killed in a road accident.

It has been said that Koko used the signs for "sad" and "cry" when her trainers told her that All Ball wouldn't be coming back. A year later she chose two more Manx kittens to care for, and named them Lipstick and Smokey.

Another gorilla, Toto, adopted in French Equatorial Africa, took a kitten named Principe under her wing, and carried the little cat around with her. Eventually Toto was sold to the circus. What happened to Principe is not recorded.

☙ WHICH ONE'S REAL? ❧

Is this Barbie come to life – and her pet leopard? Or is it Cindy Jackson, star of the TV series *Extreme Makeover*, and Cato, her wildcat hybrid?

Cindy, rock singer and doyenne of cosmetic surgery, has the dubious honor of having had more cosmetic surgery than anyone else in the world, although she reckons that many Hollywood women have had even more. "They just don't admit it."

CAT NAMES INSPIRED BY FOOD A–C		
Alfalfa	Butterbean	Cocoa
Anchovy	Butterscotch	Coconut
Bacon	Caramel	Coffee Bean
Banana	Cashew	Cookie
Bean	Cheddar	Cream
Bindhi	Cheese	Crumpet
Biscuit	Cherry Pie	Cucumber
Bread	Chicken	Cupcake
Brioche	Chilli	Cheeseball
Buckwheat	Chocolate	Cinnamon
Butter	Clementine	Custard

Her desire for all things beautiful extends to her pet, Cato, a jaw-dropping miniature leopard, with silken spotted fur and the slender muscled body of the jungle cat. He's a Bengal, a wildcat hybrid created by crossing Asian Leopard Cats with domestic cats and – like his owner's surgery – he cost A LOT.

He's named after Cato, sidekick of Inspector Clousseau in the *Pink Panther* movies, because both of them tend to pounce alarmingly out of nowhere.

Cindy's says she's finished with major surgery. From now on, it will just be minor touchups to keep her well-etched looks from withering. She supports the Cat Protection League and other animal welfare organizations. When Cato goes along with her to charity events, the audience doesn't know whom to gawk at first.

·❀ IS A CAT WORTH AS MUCH AS A GOAT? ❀·

It was in 10th-century Wales, when the value of a cat was set at four pence, or a quantity of corn, according to laws made by Hywel Dda, king of South Wales. In those days, cats may have been fairly recent imports, and were highly valued. One penny was enough to buy a lamb, a kid, a goose, or a hen, while a cock or a gander was worth twopence, and a sheep, goat or cat fourpence.

Hywel the Good called together 12 wise men, including his archbishops and nobles, and together they spent the whole of Lent fasting, praying, and creating a new set of laws. Among them was this curiosity. The amount of detail makes it seem as though it has the full weight of the law behind it. But can you imagine a stolen cat permitting itself to be dangled by the tail until it was engulfed in corn?

The worth of a kitten from the night it is (born) until it shall open its eyes is a legal penny; and from that time until it shall kill mice, two legal pence; and after it shall kill mice, four legal pence; and so it shall always remain.

The qualities of a cat are to see, to hear, to kill mice, to have her claws entire, to rear and not to devour her kittens, and if she be bought and be deficient in any of these qualities, let one third of her worth be returned.

The worth of a cat that is killed or stolen: its head is to be put downwards upon a clean, even floor, with its tail lifted upwards, and thus suspended, whilst wheat is poured about it, until the tip of its tail be covered and that is to be its worth; if the corn cannot be had, a milch sheep with her lamb and its wool is its value, if it be a cat which guards the King's barn. The worth of a common cat is four legal pence.

WHAT DID THE VICTORIANS DO WITH 200,000 MUMMIFIED CATS?

The ancient Egyptians worshipped two daughters of the sun god Ra, the cat goddesses, lion-headed Sekhmet and Bast, whose head was that of a domestic cat.

A great temple to Bast, a gentle goddess associated with both life and death, was built at Bubastis on the Nile Delta. Bast was a gentle goddess. Thousands of charming bronze statues of seated cats, some with a gold ring through the nose or ear, or a silver necklace were made as icons.

But there was also a less savory side to Bast-worship. For devotees of the goddess, offering a mummified cat to the deity was a ritual act of devotion, and for about 300 years from 600 B.C., large numbers of mummified cats were interred in special cat cemeteries, dedicated to the goddess.

A huge cat cemetery was unearthed in Egypt the late 19th century, but archaeology wasn't the super-science it is today, and no one could think what to do with almost 200,000 cat mummies. Rather ignominiously they were shipped back to the United Kingdom, pulverized and sold as fertilizer.

When cat cemeteries were discovered, only one Egyptologist was present, William Martin Conway, who wrote: "The plundering of the cemetery was a sight to see, but one had to stand well windward. The village children came [...] and provided themselves with the most attractive mummies they could find. These they took down the river bank to sell for the smallest coin to passing travelers."

Mummified cats do survive, though, and you can see some in the British Museum. The outer wrappings look like an elaborate basket weave, although often the remains inside were incomplete. Analysis of mummies shows that the feline victims were often less than 12 months old, suggesting that they were killed especially for mummification.

When a domestic cat died its owners would shave off their eyebrows as a sign of mourning. Killing a cat, even by accident, was a heinous crime which attracted the death penalty.

What do you call a cat with eight
legs that loves water?
Octopuss!

When you get home, your dog
will be pleased and lick your hand.
Cats will still be cross at you
for going out in the first place.

❧ HOW TO PREVENT CHRISTMAS CAT-ASTROPHES ❧

"Hmm, glittering baubles, lovely shiny tinsel, twinkling lights – how thoughtful! A huge new toy, just for me."

Puss can be a real pest when the tree goes up, batting at the ornaments, tugging at the tinsel and wrenching at the light cord.

Delightful as decorations are, apart from the nuisance factor when your cat is determined to dismantle them, they're also downright dangerous. Here's how to keep your cat safe at Christmas:

- Put the tree in a room that's closed off to cats. Position it off the floor, and surround with a barricade of parcels that makes it hard for kitty to leap up.
- If that's not possible, place several scat-mats (from pet shops) around the tree to try and discourage feline interest.
- Unplug lights when you leave the room. Don't leave dangling leads where a cat could chew them.
- Attach breakable ornaments to the tree with tough green plastic ties, so they are less easily dislodged.
- Put the more fragile ornaments – those that would easily break if patted over-enthusiastically – on the higher branches, out of reach.
- Don't drape tinsel where it could be dragged off the tree by a cat. It's very dangerous
  if swallowed.
- Mistletoe, holly and other evergreens and berries are poisonous to cats. Arrange them well out of reach, and remove fallen leaves and berries immediately.

❧ SHOULD CATS BE ABOVE THE LAW? ❧

In 1949, Adlai Stevenson, governor of Illinois, vetoed Senate Bill No. 93, a
provisional bill entitled "An Act to Provide Protection of Insectivorous Birds by
Restraining Cats," put forward by a well-meaning but misguided group of bird-
lovers, who had the impractical idea of making it a punishable offense for cat owners
to let their animals roam.

Stevenson's humorous good sense comes through clearly in his witty response to
the bill:

*"I cannot agree that it should be the declared public policy of Illinois that a cat visiting a
neighbor's yard or crossing the highway is a public nuisance. It is in the nature of cats to
do a certain amount of unescorted roaming. Many live with their owners in apartments
or other restricted premises, and I doubt if we want to make their every brief foray an
opportunity for a small game hunt by zealous citizens – with traps or otherwise. I am afraid
this Bill could only create discord, recrimination and enmity. Also consider the owner's
dilemma: To escort a cat abroad on a leash is against the nature of the cat, and to permit it
to venture forth for exercise unattended into a night of new dangers is against the nature of
the owner. Moreover, cats perform useful service, particularly in rural areas, in combating
rodents – work they necessarily perform alone and without regard for property lines.*
*"We are all interested in protecting certain varieties of birds. That cats destroy some birds,
I well know, but I believe this legislation would further but little the worthy cause to
which its proponents give such unselfish effort. The problem of cat versus bird is as old
as time. If we attempt to resolve it by legislation who knows but what we may be called
upon to take sides as well in the age old problems of dog versus cat, bird versus bird, or
even bird versus worm. In my opinion, the State of Illinois and its local governing bodies
already have enough to do without trying to control feline delinquency.*
*"For these reasons, and not because I love birds the less or cats the more, I veto and
withhold my approval from Senate Bill No. 93."*

❧ VERY ELDERLY CATS ❧

GLORIA

Recognized as the world's oldest cat in 2006, when she was 26, Gloria lives in Gourock, Scotland. Born in an abandoned car and rescued by Maggie McQuilkin, she lost all her teeth after a gum infection and has a diet of puréed food with an occasional treat of cooked fish.

AMBER

A Welsh stray, so her exact age was unknown. In 2005, her owner claimed to have taken Amber in, aged two or three, in 1977, making her about 30. Her cat-food diet was supplemented with treats such as hot cod, chicken and her top favorite, prawns.

SPIKE

A ginger and white tom, bought for half-a-crown in London's Brick Lane Market in 1971, Spike died at 31 years and two months in 2001. His owners, Mo and David Elkington, attributed Spike's longevity to the aloe vera gel they added to his food. At 30, Spike still hunted, had all his teeth, and ate "like a horse." His fame spread when his owners put up a banner celebrating him on their market stall in Bridport, Devon, and he used to receive fan mail from around the world, addressed to "Spike of Bridport" and "The Famous Cat, Dorset."

GRANPA

A *Guinness Book of World Records* holder, Granpa was world's longest-living cat when he died on April 1, 1998, aged 34 years, two months and four hours.

MA

A tabby, aged 34 years and one day when she died in 1957.

What happened to the cat
who drank 10 bowls of water?
He set a new lap record.

WHO BEAT ANDREW LLOYD WEBBER TO CATS?

Composer Alan Rawsthorne (1905–1971) saw the musical potential of T. S. Eliot's collection of poetry, *Practical Cats*, long before Andrew Lloyd Webber had dreamed of basing his smash hit musical, *Cats,* on the same work.

He wrote his piece, *Practical Cats,* in 1954, and recorded it with the London Philarmonia and the actor Robert Donat as narrator. The suite includes movements named after each cat, including *Bustopher Jones: the Cat about Town* (marked *Andante pomposo*), *Old Deuteronomy* (*Andante teneramente*) and the fast and furious *The Song of the Jellicles* (*Molto vivace*).

Although Alan's music was fundamentally serious, this work shows off his sense of humor and his love of cats. He once remarked that cats were easier to live with than some people.

> "*I never shall forget the indulgence with which he treated*
> *Hodge, his cat: for whom he himself used to go out and buy oysters,*
> *lest the servants having that trouble should take a dislike to the*
> *poor creature. I am, unluckily, one of those who have an antipathy to a cat,*
> *so that I am uneasy when in the room with one; and I own, I frequently*
> *suffered a good deal from the presence of this same Hodge. I recollect him*
> *one day scrambling up Dr. Johnson's breast, apparently with much*
> *satisfaction, while my friend, smiling and half-whistling,*
> *rubbed down his back, and pulled him by the tail; and when*
> *I observed he was a fine cat, saying, 'Why yes, Sir, but I have*
> *had cats whom I liked better than this;' and then as if perceiving*
> *Hodge to be out of countenance, adding, 'but he is a very*
> *fine cat, a very fine cat indeed.'*
> *This reminds me of the ludicrous account which he gave*
> *Mr. Langton, of the despicable state of a young Gentleman of good family.*
> *'Sir, when I heard of him last, he was running about town*
> *shooting cats.' And then in a sort of kindly reverie, he bethought*
> *himself of his own favorite cat, and said, 'But Hodge shan't be*
> *shot; no, no, Hodge shall not be shot.*"

From James Boswell's *The Life of Samuel Johnson* (1791)

IS CAT PLAY THE PASTIME OF AN IDLE MAN?

Or is it a rather superior form of therapy? In the 17th century, as in the 21st, time spent playing with our pets is a welcome respite from the weightier business of everyday life. Edward Topsell, a clergyman with a passion for natural history, wrote this charming observation in his book, the *Historie of Foure-Footed Beastes* in 1607.

> *It is needless to spend any time over her loving nature to man, how she flattereth by rubbing her skin against one's legs, how she whurleth with her voice, having as many tunes as turnes, for she hath one voice to beg and to complain, another to testify her delight and pleasure, another among her own kind by flattering, by hissing, by puffing, by spitting, in so much that some have thought that they have a peculiar intelligible language among themselves. Therefore how she playeth, leapeth, looketh, catcheth, tosseth with her foot, riseth up to strings held over her head, sometimes creeping, sometimes lying on the back, playing with foot, apprehending greedily anything save the hand of a man, with divers such gestical actions, it is needless to stand upon; in so much as Collins was wont to say, that being free from his studies and more urgent weighty affairs, he was not ashamed to play and sport himself with his cat, and verily it may be called an idle man's pastime.*

DO CATS THINK PEOPLE CAN SEE IN THE DARK?

Here's how readers replied, when a national newspaper in the U.K. posed this very meaningful question:

- The first question is, do cats think?
- That's what I ask our cat when I trip over him in the middle of the night.
- Scientists believe that cats do not have a "theory of mind" – they have no concept of what goes on inside the heads of any other creature. It follows that they can't

deduce what others might see, either. Cats certainly haven't worked out that people are likely to stand on them in the dark, as experience bears out.

• A lifetime of cat loving has convinced me that cats think of their own needs and nothing else. Where solipsism is concerned, cats knock celebs into a cocked hat.

WEIRD AND WONDERFUL PEDIGREE CAT NAMES

Mouser Pawprint Bigwig	*Gingertail Erik the Viking*	*Tell Tail Thomas*
Twinklepaws Tiddles	*Seablue Mini Cooper*	*Pussyfoot Golightly*
Felix Foolish Fortuna	*Burlington Queenie Fay*	*Rumbletum Greedy Guts*

❧ WHO ARE THE WHITE HOUSE CATS? ❧

GEORGE BUSH, U.S. PRESIDENT, 2001–

George Bush's pitch-black cat, known as India "Willie" Bush – or Kitty – has been the president's pet for 10 years and has her own page on the White House Web site. Named after former Texas Ranger baseball player Ruben Sierra, who was called "El Indio," she gives her address as 1600 Pennsylvania Avenue, Washington D.C., and shares the presidential residence with dogs Spotty and Barney.

India's favorite nibble is tuna-flavored kitty treats, and she enjoys hiding from her owners. If she could read, her best-loved book would be *If You Take a Mouse to the Movies*, by Laura Numeroff and Felicia Bond, and her ideal place for a snooze is under the president's bed.

In 2004, angry youths in Thiruvananthapuram, the capital of Kerala, burned an effigy of Bush in protest over the cat's name, India, saying that it was nothing but an insult to their country "because there are hundreds of thousands of Indians in U.S., and many who occupy key posts in the White House." Maybe that's why India is now referred to as Willie.

Disaffected voters can set their cat to maul a "Political Animal George W. Bush catnip toy," which comes dressed in a blue suit, complete with natty red cowboy boots, and carrying a watch which says, "Time to cut taxes"!

BILL CLINTON, U.S. PRESIDENT, 1993–2001

Socks moved in to the White House with the president in 1993 and was allowed to clamber on Clinton's shoulders, even though the man was allergic to cat dander. Feline life in the corridors of power took a distinct downturn when Buddy, a chocolate-colored Labrador Retriever, arrived in 1997. Socks spent months sulking in the basement and refused to exercise on his long lead around the White House lawns. After one cat-and-dog fight was filmed, Clinton went on TV to assure the American people that the White House pets were "making progress."

JIMMY CARTER, U.S. PRESIDENT, 1977–1981

Carter's daughter, Amy, was 10 when her father came to office, and owned several cats, including a Siamese called Misty Malarky Ying Yang.

ABRAHAM LINCOLN, U.S. PRESIDENT, 1861–1865

Lincoln's cat, Tabby, was the first feline ever to live in the White House. Lincoln had four cats while he was in office, and once remarked: "No matter how much the cats fight, there always seem to be plenty of kittens."

❧ WHAT DO CAT HATERS SAY ABOUT CATS? ❧

Just so that we can despise the writers, as much as they disliked cats ...

Try this, from the *Encyclopedia Britannica*, 3rd edition, of 1787:

> *"They are full of cunning and dissimulation; they conceal all their designs; seize every opportunity of doing mischief, and then fly from punishment. They easily take on the habits of society, but never its manners; for they have only the appearance of friendship and attachment. ... In a word, the cat is totally destitute of friendship; he thinks and acts for himself alone."*

The French poet Pierre de Ronsard (d. 1585) wrote:

> *"There is no man now living anywhere Who hates cats with a deeper hate than I; I hate their eyes, their heads, the way they stare, And when I see one come, I turn and fly."*

Hilaire Belloc, writer of *Cautionary Tales for Children*, wasn't keen, either:

> *"All that They do is venomous, and all that They think is evil, and when I take mine away (as I mean to do next week – in a basket), I shall first read in a book of statistics what is the wickedest part of London, and I shall leave It there, for I know of no one even among my neighbors quite as vile as to deserve such a gift."*

◦◦ CAN YOU TELL THE TIME BY A CAT'S EYES? ◦◦

William Salmon, 17th-century pharmacist and author of *The Compleat English Physician,* wasn't quite sure.

> *"As to its Eyes, Authors say that they shine in the Night;*
> *and see better at the full, and more dimly at the change*
> *of the Moon. Also that the Cat doth vary his Eyes with*
> *the Sun; the Pupil being round at Sunrise, and long*
> *towards the Noon, and not to be seen at all at Night,*
> *but the whole Eye shining in the darkness. These*
> *appearances of the Cat's Eyes, I am sure are true;*
> *but whether they answer to the times of Day,*
> *I have never observed."*

◦◦ HOW CAN I TELL IF MY CAT IS HAPPY? ◦◦

These are the unmistakable signs of a contented cat:

- Sleeps peacefully in favorite sunny or warm spot for many hours a day.
- Always uses the litter box or yard; never soils indoors.
- Enjoys a good scratch on his scratching post.
- Comes and goes freely, without seeming scared of the outside world.

- Is happy to be in your company, but doesn't need to be on your lap, or even in the same room, all the time.
- Likes food, and eats to satisfy his appetite.

HOW HAVE FELINES FARED UNDER THE LAW?

Over the centuries, different countries have passed all manner of inexplicable and unenforceable laws regarding cats.

The Rule of Nuns was an early 13th-century edict under which nuns were forbidden to keep any animal other than a cat. A hundred years earlier, nuns were banned from wearing any animal skin that was more expensive than that of a lamb, or a cat.

In the Middle Ages animals were commonly accused of crimes, or expected to appear in court as witnesses at the trials of murderers and thieves. In the 16th century, Bartholomew Chassenée explained to the French court that his clients, some rats, hadn't turned up for their court appearance in Autun, because it was too dangerous for them to cross the cat country that lay between their home and the courthouse.

In pre-medieval Wales, the legal definition of a hamlet was a place with:

Nine buildings
One plough
One kiln
One churn
One bull
One cock
One herdsman …
… One cat

Under the same Welsh laws, if a husband and wife were to separate, and owned only one cat, it would go to the husband. If there were others, then they would be taken by the wife.

Ship your corn by sea, and you're not covered for damage by rats – unless you can prove that the ship went to sea without a cat. If the captain was so foolish as to set sail without a ratter on board, marine insurance allows the owners of the goods carried to claim damages.

☙ WHY SHOULD MY CAT BE VACCINATED? ☙

Just like their owners, cats can pick up nasty bugs very easily.

The three main viruses that attack cats are feline leukemia virus (FeLV), feline enteritis and cat flu. If unvaccinated, your pet can pick these up from other cats, from your shoes or clothing, from litter boxes or food dishes, or simply from the ground. Mixing with other cats, either at home or in a boarding facility, increases the risk, but if your cat lives indoors, it will have less natural protection from any source of infection you unwittingly bring into the house. All cats should be vaccinated to protect them.

Your cat should have its first vaccination between nine and 12 weeks, with a follow-up shot two weeks later. After that, an annual booster is all that's needed to keep the protection going. One vaccine covers all three diseases.

Feline Leukemia

The biggest threat to young cats after traffic accidents, feline leukemia lowers immunity, and can produce cancers. It's spread in feline saliva, and transmitted if an infected cat licks or bites another. It can take years for symptoms to appear after infection.

Feline Enteritis

Highly contagious, this disease can spread rapidly via clothing, shoes, litter boxes and food bowls. Symptoms include exhaustion, loss of appetite, vomiting and diarrhea, and one cat in 10 who's infected will die.

Cat Flu

Two viruses combine to cause this nasty disease, which although not as deadly is still very infectious and can leave its victims permanently damaged. Symptoms include coughing and sneezing, runny eyes and nose, lassitude, fever and exhaustion.

How very unpleasant. Do your cat – and other cats – a good turn, and always keep her vaccinations up to date.

☙ WHERE DO CATS SLEEP? ☙

A few favorite places …

• On top of the fridge – it's that lovely warm draft coming up the back.

- In the very middle of the path from the living room to the kitchen.
- Snuggled in between two humans, especially if they were planning on cozying up together.
- Crammed underneath a desk lamp. The smell of scorching fur gives the game away.
- On any pile of papers, carrier bag or seat that their owners need.
- On your pillow.
- In an open cabinet.
- Under the oven.
- Anywhere that isn't their bed.

⊛⊛ WHY HAS MY CAT'S SLEEPING PATTERN CHANGED? ⊛⊛

If there's a marked difference that lasts for more than a day or two, take your cat for a health check.

A dopey and lethargic cat may be ill. Conversely, sleeping less than usual could indicate a thyroid problem, which is not uncommon in older cats.

CAT'S FAVORITE CHRISTMAS SONGS
Wreck the Halls
Joy to the Curled
The First Meow
Silent Mice
Fluffy, the Snowman
Jingle Balls
Have Yourself a Furry Little Christmas

⊛⊛ WHY IS MY CAT'S TONGUE SO ROUGH? ⊛⊛

If you've ever been treated to an affectionate licking, you'll know just how rasping and coarse Muffin's tongue can feel. The reason is that a cat's tongue is its built-in comb, and is thickly coated with tiny papillae. These little hooks are designed to rake efficiently through the cat's coat to remove tangles and dirt. They're also perfect for cleaning the last scraps of meat off bones, and are pretty good at cleaning out a cat bowl until it's so shiny that so you'd never guess it was full just 10 minutes ago.

⋅❀ WHAT MAKES A PEDIGREE A PEDIGREE? ❀⋅

In the late 19th century, the first Cat Clubs in the United Kingdom and United States were formed, in order to verify feline pedigrees, and create standards for different breeds. Now there are registries in most countries, and many have more than one.

The whole business is wildly inconsistent. What counts as a pedigree in one country isn't recognized in another, because the different registries not only admit different breeds, but a breed that's known by one name in one country may be registered under a different breed type in another.

There is lively debate, to put it politely, in the cat-breeding world, about the best way to approach the question of pedigree. Cat-breeding politics center on two opposing standpoints. Experimental breeders love putting together cats of different colors, coat types and species, to create a proliferation of new cat types. Traditionalists would rather stick to a smaller range of breeds, and keep each one distinct, with limited variations.

A century or so ago, most cats came from old breeds which had emerged through natural inter-breeding. But in the last hundred years, breeders have created new breeds, like the Tonkinese and Angoras, from scratch. As it were.

Some of these are straightforward longhaired versions of an existing shorthaired breed. Others work on an existing mutation, such as a wavy coat, which would die out naturally in the wild, and, through careful breeding, replicate that characteristic to create a new breed.

Taken to extremes, this kind of breeding can create very strange creatures, like the furless Sphynx, or the pocket-sized Munchkin, which some registries refuse to recognize. Looking at these peculiar cats that seem so un-catlike begs the question of how much selective breeding is too much.

❧ WHO WAS A FELINE MARRIAGE BROKER? ❧

Diminutive French artist Henri de Toulouse-Lautrec (1864–1901) drew a poster featuring the Irish singer May Belfort in which she is shown wearing an outsized red baby outfit and carrying a tiny black cat – the costume she wore to sing the music-hall song "Daddy Wouldn't Buy Me a Bow-wow," which includes the lines:

I love my little cat, I do
With soft black silky hair
It comes with me each day to school
And sits upon the chair
When teacher says, "Why do you bring
That little pet of yours?"
I tell her that I bring my cat
Along with me because …

Daddy wouldn't buy me a bow-wow, bow-wow,
Daddy wouldn't buy me a bow-wow, bow-wow,
I've got a little cat,
And I'm very fond of that –
But I'd rather have a bow-wow-wow!

May Belfort was a cat lover even when not dressed up in a size XLL romper suit, and evidently asked Lautrec to try to find a suitable mate for her cat. He wrote to a friend: "Miss Belfort wants a husband for her cat. Is your Siamese man for the job? Let me know, and let's fix up a rendezvous."

❦ WHO LETS THE CAT OUT? ❦

Not much is known about Benjamin Franklin King (1857–1894), who's probably best remembered for these philosophical lines from his poem, "The Pessimist."

Nothing to do but work,
Nothing to eat but food,
Nothing to wear but clothes
To keep one from going nude.

It's safe to assume, though, that there was a cat in King's household. No one who without first-hand experience of a cat's overwhelming desire to come in – or go out – could have written this poem:

That Cat

The cat that comes to my window sill
When the moon looks cold and the night is still –
He comes in a frenzied state alone
With a tail that stands like a pine tree cone,
And says: "I have finished my evening lark,
And I think I can hear a hound dog bark.
My whiskers are froze, 'nd stuck to my chin.
I do wish you would git up and let me in."
That cat gits in.

But if in the solitude of the night
He doesn't appear to be feeling right,
And rises and stretches and seeks the floor,
And some remote corner he would explore,
And doesn't feel satisfied just because
There's no good spot for to sharpen his claws,
And meows and canters uneasy about
Beyond the least shadow of any doubt
That cat gits out.

CAT NAMES INSPIRED BY FOOD D–M

Dahl	Hickory Bean	Kiwi
Date	Honey	Lemon
Dinner	Honey Bun	Lambchop
Dumpling	Honeydew Melon	Licorice
Figgy Pudding	Hot Dog	Macaroni
Fondue	Huckleberry	Mackerel
Fudge	Humbug	Marmalade
Ginger	Icing	Marmite
Gingerbread Boy	Jam Sandwiches	Marshmallow
Gingersnap	Jellybean	Mr. Chutney
Gravy	Juniper	Mr. Peanut
Haggis	Kebab	Muffin
Halva	Kipper	Mushroom
Hazelnut	KitKat	Mustard Seed

•⁰• WHEN DOES A LOST-AND-FOUND POSTER BECOME A WORK OF ART? •⁰•

Tracey Emin loves Docket. Official.

> *Every morning I wake up and he's there, little paws,*
> *little ears. He mutters a few meows, rolls over on his*
> *back and smiles. And I think, "How could anything so*
> *perfect have ever been created?"*

The notorious conceptual artist behind *My Bed* – which contained a whole lot of Emin's life detritus, but no Docket droppings – heard the news that she'd been made a member of the prestigious Royal Academy while driving her cat to the vet. "I couldn't talk because I was driving, but the phone call was from Gary Hume RA asking me, if I were given the honor of being a Royal Academician, would I accept it? As I drove down the Mile End Road I shouted across the car: 'Of course I will! Of course I will!'"

Docket appears in Emin's work – he's shown on a travel card wallet she designed to mark the 60th anniversary of the U.K. Arts Council, with the caption, "We've got fur and lots of ears."

Docket has pedigree leanings – he's half mog, half British Blue, and he hails from Vallant Road, where gangland bad boys, the Kray twins, came from. But Docket definitely isn't a hard man. He loves a cuddle, is very chatty and even follows commands, occasionally.

His best friends are Fleabite – greasy fur and no teeth – and ginger Pious, who's rather camp and lives a few doors down.

When Docket went missing, Emin did what every urban cat owner would do. She put up posters all around the neighborhood, asking people to keep a look out for him. The art cognoscenti, scenting another Emin-installation, were quick to respond. Posters were ripped down as fast as she could put them up, and were soon changing hands for up to £500, despite her protest that, look, she really had just lost her cat. He turned up again.

CATS AND COMPUTERS 7
It's fun behind the tower – you can unplug
all the USBs with a couple of paw flicks.

❦ WHAT'S MISSING FROM AUSTRALIA? ❦

It's the only continent that doesn't have an indigenous species of wild cat.

There are numerous species of wild cat inhabiting the forests, plains and jungles of the world. Most live by themselves, except for lions, which live in prides, and they hunt a variety of prey depending on what smaller creatures live locally.

Almost all types sport glorious camouflage-coats, stippled with spots and bars of color that let them merge silently into the light and shade of their natural surroundings. Forest-dwelling wild cats may even have black or very dark coats, so they can glide stealthily through the shadowy undergrowth. This inherent beauty makes cats vulnerable to hunters, and although there are international laws to protect them, wild cats are often the target of poachers.

❧ WHICH BOWLS ARE BEST? ❧

As on so many important topics, cats have an opinion on feeding bowls. According to a spokes-cat, the optimum bowl is:

- Shallow. I prefer my whiskers to be splayed, not squeezed.
- Cleaned thoroughly before every meal. No dried-on bits thank you.
- Ceramic, not plastic. I am very fussy about residual whiffs of dishwashing liquid, or old food. Plastic is seldom pristine enough for me.
- Individual. Don't make me share with my brother, no matter how much we love each other. He wolfs all the food, greedy hog, and I'm left hungry. Put our food down at the same time, but give us a bowl each, please.

❧ HELP YOUR CAT TO CLIMB ❧

Kosh is designed to climb. He's got the paws, the claws and the strength for it, and in the wild it's what he just loves to do.

Climbing gives him exercise, and a lofty perch makes him feel safe. So, how can you make your living room resemble the scrubby woodland that Kosh hankers for?

Take a trip to the pet superstore and you'll find an array of scratching posts and climbing stations with built-in platforms, enclosures and climbing frames. They're not beautiful, but they're highly feline-friendly. So if you've got the space, and aren't bothered about the esthetics, your problem's solved.

Do you have a tall cabinet or wardrobe which might just provide a desirable roost? When Kosh's back is turned, you can move other furniture to give him a route to the top. He'll be more likely to give this a try if you don't point it out to him. Cats like to discover things for themselves, thank you very much.

If you're handy with saw and screwdriver, put up a few high shelves especially for him, or clear a high shelf of books to give him a bit of space. Any shelf needs a non-slip surface, such as a bit of well-secured carpet. Otherwise Kosh could whoosh straight off the end the first time he takes a flying leap.

He'll love to scrabble up a length of carpet or cloth-covered board that you've attached to a wall. Position an accessible surface somewhere near the top, otherwise his only way down will be by letting go.

The sturdy cardboard inner tube from a roll of carpet makes a wonderful virtual tree. Cover it with carpet and secure from ceiling or wall.

❧ HOW MANY MICE? ❧

These working cats repaid their employers time and again, and applied themselves vigorously to the job of keeping vermin down.

- Deadly tabby Towser prowled the granaries of a Scottish distillery for 23 years, during which time she sent more than 28,000 mice to meet their maker. A statue was put up in the distillery grounds in her honor.
- Minnie was ratter-in-residence at a London sports stadium for six years, where she dispatched an average of 2,000 rats each year.

*It is a very inconvenient habit of kittens
(Alice had once made the remark) that,
whatever you say to them they always purr.
'If they would only purr for "yes"
and mew for "no," or any rule
of that sort,' she had said, 'so that
one could keep up a conversation!
But how CAN you talk with a person
if they always say the same thing?*

Alice, *Through the Looking Glass* by Lewis Carroll

✦ WHO'S GOING UP ... AND COMING DOWN? ✦

- Andy, pet of a Florida state senator, survived a fall of 200 feet.
- In 1950, a four-month-old kitten is said to have trailed its owner to the peak of the Matterhorn, all 14,693 feet of it.
- Mincho, an Argentinian cat, went up a tree – and never came down again. She remained in her roost for six years, presumably living off passing birds.

THINGS PEOPLE ASK...

... when they ring the cat care help line.
How many calories are there in a mouse?
I've just bought a one-year-old neutered tom.
How long will it be until I can breed from him?

✦ FELINE MUSICAL MISCELLANY ✦

CAT'S CHOIR

In "La Gata I en Belitre," a Catalan folk song arranged by Francesc Pujol, the male chorus is called upon to imitate the sound of cats mewing.

TAILS OF THE ORGAN LOFT

A black cat that lived in the church of St. Clement Danes used to doze on top of the organ pipes – until the organist began playing. The shriveled remains of another less fortunate creature were found clogging one of the long organ pipes in Westminster Abbey. So that's why it was sounding so flat!

STRANGE BEQUESTS

An elderly American cat lover who died in the 1870s left money to establish a cats' hospital. He instructed that an accordion "be played in the auditorium by one of the regular nurses, to be selected for that purpose exclusively, the playing to be kept up for ever and ever, without cessation day and night, in order that the cats may have the privilege of always hearing and enjoying that instrument which is the nearest approach to the human voice."

A Sweet Serenade

As a child, Jenny Lind, the opera singer known as the "Swedish Nightingale," used to sing to her cat. She would sit in the window street, watching the people go past on their way to church and singing sweetly to her feline audience, at which "the people passing in the street used to hear and wonder."

The Gut of a Cat?

Whoever said: "No sounds are so captivating as those made by the men of sin who rub the hair of the horse across the bowels of the cat" clearly didn't know that "catgut" violin strings are – and always have been – made from the entrails of lambs.

WHAT WAS RUMPEL REALLY CALLED?

A titled cat! How appropriate for the pet of poet laureate Robert Southey (1774–1843), fellow Lake poet to William Wordsworth. He wrote to a friend:

> *Alas, this day poor Rumpel was found dead, after*
> *as long and happy a life as cat could wish for, if cats*
> *form wishes on that subject. His full titles were:*
> *The Most Noble, the Archduke Rumpelstiltzchen,*
> *Marcus Macbum, Earl Tomlefnagne, Baron Raticide,*
> *Waowhler and Scratch. … I believe we are each and*
> *all, servants included, more sorry for his loss, or, rather,*
> *more affected by it, than any of us would like to confess.*

CAT NAMES INSPIRED BY FOOD N–R		
Noodle	Passionfruit	Porkchop
Nutmeg	Peanut	Prawn
Oatmeal	Pepper	Pudding
Okra	Peppermint	Pumpernickel
Oregano	Pepperoni	Pumpkin
Pancake	Pickles	Raisin
Parsley	Pie	Ravioli
Parsnip	Pilchard	Rhubarb

WHO NAMED HER CAT AFTER A STREET?

Sister of the colorful genius Augustus John and lover of celebrated sculptor Auguste Rodin, Gwen John (1876–1939) was herself quiet, gentle and reclusive. Her subdued paintings are the antithesis to all that flamboyant testosterone, and are beautiful for their subtle play of light and shade and restrained palette.

She used her subjects as objects and considered a cat and a man to be equal in this respect. Yet her drawings show a keen perception of the way cats use their bodies, and capture the essence of what makes a feline a feline. Even so, despite her fierce attachment to her cats, they only appear in six paintings and a number of drawings and watercolors. Probably they weren't the most cooperative of sitters.

In 1904, leaving Wales for Paris, Gwen John settled at 19 Boulevard Edgar Quinet, a somber street lined with funerary shops. She quickly found herself a white-chested tortoiseshell cat and named her Edgar Quinet after the street. This cat appeared in evocative pencil drawings, often enhanced with a simple monochrome wash, including "Wide-Awake Tortoiseshell Cat" and "Napping Tortoiseshell Cat." Gwen John beautifully captures the most fleeting gesture, the licking of a paw, the total relaxation of sleep.

Living alone made Gwen particularly dependent on Edgar Quinet for companionship, and she thought very highly of her, perhaps too highly.

Her fears were realized when Edgar Quinet went missing twice and, the second time, failed to return. Gwen was heartbroken, and went out at night searching for her pet. Some months later, a friend offered to give her a kitten, but she refused, not wanting to become so attached to a cat again and instead preferring to put the energy of loving a cat into her drawings.

Nonetheless, she did have other cats. Ceridwen Lloyd-Morgan, who edited Gwen's letters and notebooks, published in 2004, struggled to transcribe the artist's jottings, but noted that the faded pencil marks on rough paper that had darkened with age and were sometimes obscured by marks and cat vomit left little contrast between the paper and writing.

As she grew older, Gwen John became more and more of a recluse, and began to neglect her health. In 1939, aged 63, she traveled to Dieppe, France, where she collapsed and died. Her brother noted that, on this last journey, while she hadn't brought any baggage, she had, of course, thought to make provision for her cats.

❀ BEAM ME UP, SPOTTIE? ❀

Star Trekkies won't need telling that Spot was a cat belonging to Data, who was not only chief operations officer on the starship *Enterprise*, but also a sentient being designed to look just like a human – otherwise known as an android.

In the series, Spot survived being mutated into an iguana, and back again. He only really got along with Data, and treated the rest of the crew with disdain.

Spot started out as a male Somali cat, but later was seen as a female orange tabby, and eventually gave birth to a litter of kittens.

Androids don't do emotion, but even so, Data formed a deep connection with his pet, and composed a heartfelt poem to him/her which included these words:

O Spot, the complex levels of behavior you display
connote a fairly well-developed cognitive array.
And though you are not sentient, Spot, and do not comprehend,
I nonetheless consider you a true and valued friend.

⚬☙ WHEN SHOULD I CALL THE VET? ☙⚬

It can be hard to tell when that master of subtlety – a cat – is unwell. Most felines abhor making a fuss when there's something wrong. It's part of their nature that harks back to life in the wild, when predators would pick off any animal that was less than 100 percent fit.

Keep a close eye on your pet, and if you do spot symptoms, even slight, don't delay in making a visit to the vet. Some serious complaints present only mild signs at the outset, so it's always better to be safe, rather than sorry.

CHANGES IN APPETITE FOR FOOD/DRINK

Don't worry if Pipkin refuses food or drink for a day or two. She might have gotten tired of kipper-with-broccoli and want a change, or have begged, found or stolen a dinner elsewhere. Longer appetite loss could indicate any of a whole range of problems. Get her checked over. A greatly increased appetite or thirst indicates problems with liver, kidney or thyroid. Crying for food but refusing it when offered is also a cause for concern, and could indicate a serious underlying condition.

WEIGHT LOSS

Losing weight is usually sign that something's amiss, so take your cat to the vet. Older cats often become very skinny, and it's not a problem unless the weight loss is very sudden. If Pipkin's still eating well, she may need treatment for kidney problems or overactive thyroid.

BAD BREATH

Inflamed gums are usually at the root of foul breath. Bacteria gather where tooth meets gum and cause halitosis. Feline flu virus can cause severe inflammation, which leads to bad breath and drooling, so if your cat isn't inoculated against it (and she should be), see the vet. Give her crunchy foods, ideally non-splintering bones, to keep her teeth and gums in good shape.

VOMITING

The occasional up-chuck or regurgitated hair ball is par for the course for all cats. Try providing grass to nibble to help with hairballs. Abnormal, persistent vomiting is a cause for concern.

Diarrhea

If caused simply by eating something undesirable, it will pass quickly. Diarrhea can also be a symptom of infections, parasites or other conditions. Seek advice if persistent.

Urinary Problems

Keep an eye on how often your cat uses the litter box. Both increase and decrease in urination should be checked out, as there may be a blockage or other serious condition.

Excessive Scratching or Licking

Fleas or ticks are the usual suspects if your cat starts scratching manically, and are readily treated with preparations available from your vet. Allergy is another possible cause. If a cat is licking more than usual, or is concentrating on a particular spot, investigate. Cats always lick wounds thoroughly, so the coat may be covering a cut or sore that needs attention. Ill health may show in the coat, so if your cat's fur looks dull or staring, rather than smooth and glossy, seek advice.

EYE PROBLEMS

Various infections cause rheumy, inflamed eyes, most of them highly infectious. Your vet will prescribe antibiotics. Keep your cat vaccinated to avoid infections. If the inner eyelid doesn't retract as normal, it's a sign that your cat is ailing.

EAR PROBLEMS

Ear mites are irritating little parasites that cause infections and will have Pipkin shaking her head and pawing at her ears. A build-up of dark, gritty wax is another sign. Your vet can give suitable medications.

SNUFFLES, RUNNY NOSE AND SNEEZING

Allergic reaction or a foreign body such as a grass blade in the nose is a possible cause. Red eyes and runny nose can indicate cat flu. Isolate your cat, and see a vet immediately. Avoid infection by making sure Pipkin is vaccinated.

COUGHING

Cats can get a bad chest for a variety of reasons, from a mild treatable infection, to more serious ailments of chest and heart. Get your vet to diagnose and treat appropriately.

UNSTEADINESS, FALLING

Problems with balance often stem from injuries or middle-ear infection. Even if your cat appears otherwise well, always consult a vet if she develops a staggering walk.

BEHAVIORAL CHANGES

Listlessness, soiling, aggression, hiding, excessive sleeping or any other unusual behavior can all be subtle indications that something's amiss. Consult your vet.

NEWLY NEUTERED
A tomcat was running back and forth along the back alleyway for hours. A neighbor rang his owner and asked what was going on. The owner said, "I had him neutered today, and he's cancelling all his dates."

✦❧ IF ONLY THEY COULD TALK ❧✦

Hector Hugh Monro, better known as the ace Edwardian short-story writer Saki (1870-1916), toyed with the idea of just what would happen if cats could speak, in his wickedly witty tale, "Tobermory."

The scene is a house party, where one of the guests has taught Lord and Lady Blemley's cat, Tobermory, to talk. It quickly becomes apparent, however, that Tobermory pays no regard to the usual social niceties.

> *"What do you think of human intelligence?" asked Mavis Pellington lamely.*
>
> *"Of whose intelligence in particular?" asked Tobermory coldly.*
>
> *"Oh, well, mine for instance," said Mavis with a feeble laugh.*
>
> *"You put me in an embarrassing position," said Tobermory, whose tone and attitude certainly did not suggest a shred of embarrassment. "When your inclusion in this house-party was suggested Sir Wilfrid protested that you were the most brainless woman of his acquaintance, and that there was a wide distinction between hospitality and the care of the feeble-minded."*

Nor does Tobermory have any qualms over revealing what he's learned from his nightly perambulations along the ornamental balustrade outside the bedrooms.

> *Major Barfield plunged in.*
>
> *"How about your carryings-on with the tortoise-shell puss up at the stables, eh?"*
>
> *The moment he had said it everyone realized the blunder.*
>
> *"One does not usually discuss these matters in public," said Tobermory frigidly. "From a slight observation of your ways since you've been in this house I should imagine you'd find it inconvenient if I were to shift the conversation to your own little affairs."*

The panic which ensued was not confined to the Major. The party wastes no time in plotting to silence the over-talkative cat, and prepare a plate of fish laced with strychnine. But before the poison can be administered, Tobermory perishes at the claws of the rectory tom, and the guests breathe a huge sigh of relief, as their secrets have gone with him to the grave.

CAT NAMES INSPIRED BY FOOD S–W

Saffron	Taffy	Tuna Can
Salt	Tangerine	Turkey
Sausage	Tapioca	Turnip
Sesame	Toast	Twiglet
Smartie	Toffee	Twix
Spice	Tomato	Vanilla
Squid	Treacle	Veggie
Sugar	Truffles	Vinegar
Sugar Pie	Tuna	Waffles

❧ WHEN A CAT WANTS TO MOVE IN ❧

He's often there in the garden. When you go out, he purrs, and rubs himself around your legs. He looks a bit skinny. Perhaps you offer a bowl of milk. Lately he's been hovering around the door, or maybe he's managed to get into your house, and you've found him exploring.

When a strange cat shows all the signs of wanting to move in with you, don't assume it's an unloved stray, unless it shows obvious signs of neglect. Owners would deny it, but cats can be fickle creatures, who'll develop a passionate cupboard love for anyone who puts out a few enticing tidbits. So before you give Pookie a name and offer him the run of your home, remember that someone else might be pining for their Albert, and make a few simple checks.

- Does he have a collar with a name and address tag? Pop around and see the owners, and alert them to their pet's tricks. A little addition to his breakfast bowl might be all they need to give him, to make sure he comes home to them every night.
- Ask around locally to see if anyone has a cat that fits your visitor's description.
- Check for "lost cat" ads around your neighborhood, in local shops or in the newspaper's "lost and found" column. Consider putting an ad up yourself.
- If the cat is making a real bid to move in, and has been around for a week or more, take him to a vet, who can scan him to see if he's micro-chipped. The vet can also confirm a cat's gender, make a guess at its age and tell you about its general condition, including whether or not it's been neutered.

WHEN CAN I LET MY NEW KITTEN OUTSIDE?

Kitty should stay inside until she's been neutered and had her vaccinations, which should all be done by the time she's around six months old. Assuming that she's lived in your home for some weeks by this stage, and well used to her surroundings, you can safely let her explore the great outdoors.

To start with go outside with her, and keep an eye on where she roams. Kittens have different personalities, and if yours is an extrovert, she'll be off and over the fence fairly soon. But she'll have marked her patch with her droppings and scent glands first, so she'll be able to sniff her way home. In the early days, don't leave your kitten unattended outside for too long at a time, and bring her home by calling, then rewarding her with petting and perhaps her favorite snack.

 I love cats because I enjoy my home,
and little by little,
they become its visible soul.
Jean Cocteau, French poet and filmmaker

❧ WHY DOES MY CAT BITE? ☙

Cats don't bite people nearly as often as dogs do, but if they do sink their gnashers in, chances are you'll end up with an infected wound. It may be bacteria on the cat's teeth that are to blame, but certainly cat bites, whether inflicted on humans or on other cats, have a tendency to turn nasty.

It's very unusual for a cat to give an unprovoked bite, even if it's a bit tricky for owners to work out exactly what counts as provocation. Millie could be sitting purring on your lap enjoying a stroke one moment, and the next she's suddenly given you a nip. There could be a conflict in the feline mind between the delight of being petted, and the dismay of having an intrusion into your personal state. A cat can put up with – even relish – so much fondling. But suddenly, enough's enough, and she puts a stop to the sensation in the only way she knows how. Most cats will give some kind of warning that they're reaching the end of their tolerance, so watch out for a twitching or waving tail, a low growl, or a feinted bite and quickly put her down.

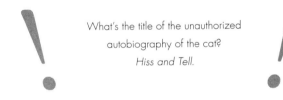

What's the title of the unauthorized autobiography of the cat?
Hiss and Tell.

❧ DO YOU, WOOKIE, TAKE TREACLE, TO BE YOUR LAWFUL WEDDED …? ☙

No, it's not legal, and you could argue that it's somewhat pointless, but there's no reason why cats can't "get married." Should your pet wish to take the plunge, rest assured that you can find feline wedding veils, top hat and tails, ruffled neckwear for the bridesmaids and other bridal apparel on the Internet.

Here are the tales of two happy cat-couples.

Phet and Ploy, two rare diamond-eyed cats, wore pink for their wedding ceremony in a Thai discotheque in 1996. The party cost more than 400,000 Thai bhat, or $12,400 at today's prices.

The courtship of Armand and Cassie, reads like a game of feline consequences.
Armand, a hairless Sphynx

met

Cassie, a furry tortie

at

a California Petsmart when they were having their photos taken.
He said to her:
"May I sniff your nose?"
She said to him:
"My owner thinks it would be cute if we got married."

The consequence was:

They were carried down the Petsmart aisle, while a flutist played the theme from
The Pink Panther. Armand, resplendent in black satin and velvet waistcoat and top
hat, exchanged vows, via his owner, with Cassie, who wore a white gown with gold
California poppy decoration. The couple exchanged collars – Cassie's was pink with
her name spelled out in crystals. After the wedding, the newlyweds had their picture
taken with Santa Claus.

And the world said:

"That's surreal."

·❀· BIBS SAVE BIRDS ·❀·

Researchers in Western Australia had the weird but effective idea of fitting predatory cats with brightly-colored clip-on bibs in an attempt to curb their hunting. The bibs didn't interfere with the pets' everyday lives and they could still run, jump, wash themselves and eat normally. But the dash of brilliant color was enough to warn prey of the cats' approach, and the bibs themselves hindered the stealthy creeping, stalking and pouncing that spelled death for so many small creatures.

Over 80 percent of the bird-catching cats had their activities hampered, and the bibs also cut down on the numbers of small mammals, frogs, newts and other creatures the cats had been catching before being bibbed.

THINGS PEOPLE ASK...
How can I stop my cat stealing
my husband's toothbrush?
My cat poo'ed on the rug and the poop's
stuck in the vacuum cleaner. What do I do?

UNTIL WE MEET AT THE RAINBOW BRIDGE...

If only cats lived as long as humans. But alas, they don't, and so it's inevitable that over the years, cat owners have to steel themselves to saying a last "goodbye" to a succession of beloved pets. One idea that has proliferated since the birth of the Internet, and has given solace to thousands of bereaved pet owners, is The Rainbow Bridge. The bridge appears in an anonymous poem which was possibly inspired by an ancient Norse legend of a rainbow bridge called Bifrost, used by the gods to go between Valhalla and Earth.

In the modern version, The Rainbow Bridge is a place "just this side of heaven," where pets that have died – cats, dogs, rabbits, birds – live on in youthful good health, playing in green fields and resting in the sunshine, waiting for their owners to travel the same path and join them. When at last each animal hears that longed-for familiar step, it runs to meet its owner and, reunited, together they cross The Rainbow Bridge.

Boo hoo! The poem is a real tear-jerker, but at the same time, the loss of a much loved cat who has been a friend, companion, comforter and source of delight is painful and hard, and The Rainbow Bridge gives comfort and hope to many.

When a cat dies, you could choose to have him or her buried at a pet cemetery, or you can find a peaceful spot in your garden and bury your pet there. Make sure the grave is quite deep, and place a stone or other heavy object over the spot, to prevent foxes from excavating.

It takes time to get over the death of a cat that has been a loyal companion for many years. There's bound to be a gap, and it's natural to grieve. Try to focus on the happy times you had together. Some owners like to remember their pets by keeping photos on display, or creating a "memory book" with written thoughts and pictures. Another positive way to mark your cat's memory is by donating money, or a regular hour or two of your time, to a local animal rescue charity. What better way to celebrate the wonderful and very special relationship you've enjoyed with your cat?

WHAT IS A SQUITTEN?

No, it isn't possible for a cat to mate with a squirrel or, for that matter, a kangaroo, a rat or a rabbit. Every so often though, a cat crops up with a deformity or genetic glitch that changes its appearance, so that it looks a bit like a hybrid between a cat and some other creature. Cats have been reported with foreshortened front legs, coupled with a feathery longhaired tail and an opposed front toe giving the ability to grip, and they certainly do look a little squirrelish. But however much like a squirrel a kitten appears, the two species are genetically incompatible and a "squitten" is a physiological impossibility.

CAN CATS HAVE WINGS?

It seems cats can, on occasion, be winged but, surprise – they can't fly! The appendages that, at a quick glance, look like wings, and which may even flap up and down when the cat runs, are usually flaps of matted fur which can develop if long-haired cats aren't properly groomed, and drop off when the cat moults. Other possible sources of "wings" are skeletal deformities and diseases of the skin.

An early 19th-century "winged cat" was described by American author and naturalist Henry David Thoreau, who wrote:

"A few years before I lived in the woods there was what was called a "winged cat" in one of the farmhouses in Lincoln nearest the pond. ... her mistress told me that she ... was of a dark brownish-gray color, with a white spot on her throat, and white feet, and had a large bushy tail like a fox; that in the winter the fur grew thick and flattened out along her sides, forming strips 10 or 12 inches long by two-and-a-half wide, and under her chin like a muff, the upper side loose, the under matted like felt, and in the spring these appendages dropped off. They gave me a pair of her 'wings,' which I keep still."

An even more curious winged specimen was reported in Derbyshire in 1897, where the local paper had this to say:

"Mr. Roper of Winster ... shot what he thought to be a fox. ... It proved to be an extraordinarily large tomcat, tortoiseshell in color with fur two-and-a-half inches long, with the remarkable addition of fully-grown pheasant wings projecting from each side of its fourth rib. ... Never has its like been seen before, and eyewitnesses state that when running, the animal used its wings outstretched to help it over the surface of the ground, which it covered at a tremendous pace."

The truth behind this tale will never be known, because unfortunately no one thought to preserve the creature, which began to rot in the summer heat and was hastily buried.

In the early 1930s, a winged cat was caught at a house in Oxford, England, and was displayed at Oxford Zoo for some time afterwards. Its wings were six inches long and grew out of the body, just in front of the hindquarters.

❧ WHY ARE THERE SO MANY CAT SUPERSTITIONS? ❧

It could be because cats go out at night and therefore must be mixing with goblins, demons and other creatures that shun the light. Black cats are darkly sinister, and their color has long been associated with evil. Just as people who set themselves apart from society and lived alone were feared, so their pets were tarred with the same brush. In medieval times, the cat was thought of as an occult "familiar," and people believed witches could turn themselves into cats at will.

Add to that the spooky glitter of a cat's eyes in the night, the demonic yowl of a cat in heat, and the way cats always seem to stroll away from trouble, and you begin to understand why they've inspired dread and awe over the centuries.

❧ WHEN A CAT GOES MISSING ❧

It's a horrible, sinking feeling, when you suddenly realize your cat hasn't come back at her usual time. She's missed a meal or two, and you can't quite remember when you last saw her. If the hours go by, you've tried calling, and you're starting to feel really worried, what should you do?

- Keep calling your cat at regular intervals. Try hourly, or even more frequently, once you realize that she's missing. The best time is at dusk and into the night, when it's quieter and cats are on the alert. First try her usual haunts, then spread the search further afield. Call in a gentle voice, and pause and listen for up to a minute between each call. It could take your cat that long to recognize and respond to your voice. If she's trapped somewhere, or injured she may not be able to come when you call, but might "meow" her reply. Continue calling regularly. Cats have been known to reappear after days or even weeks of absence.
- Rattle her food bowl on the ground, or bang a spoon on the cans. Any signal that she might recognize and associate with food. Leave out some strong-smelling morsels of food to attract her.
- When you search after dark, shine a torch up into the trees and dark spots where she might be stuck. A frightened cat may be too terrified to respond vocally to a call, even if she can hear you.
- Don't stand guard, waiting for your cat to reappear, as this may deter her.
- Search your own outbuildings, and ask neighbors to look in garages, sheds, greenhouses and any other places where a cat could be trapped.
- Road traffic is a menace to cats. Call your local vet and see if any injured cats have been brought in. Your local government may have a policy for dealing with dead cats found on the highway. Your vet may be able to advise on this.

- You can also take on the task of searching around roadways yourself. Harden yourself to look in gutters, ditches and on the tops of hedges, where a cat's body could be flung by an impact. Horrible indeed, but it's far better to know what has happened to your pet, even if it is the worst.
- Take comfort from the thought that most cats do turn up unharmed, and are none the worse for their adventures, although whether the same can be said of their owners' nerves is another question.
- Above all, make sure your cat is both micro-chipped and neutered or spayed. Male cats are far more likely to roam if they're unneutered, and females who are not spayed run a real risk of coming home pregnant if they stray while in heat. It's a myth that it's good for females to have at least one litter. It isn't, and she will stay kitten-like herself for much longer if she's spayed when young. If you'd love to own a kitten, then please consider homing one of the thousands of unwanted baby cats that are found abandoned and left to starve every year. Your local animal shelter will be delighted to help.

*Cats are intended to teach us that
not everything in nature has a purpose.*
Garrison Keillor,
author of *Lake Wobegone Days*

HOW CAN I REHOME MY CAT?

Sometimes circumstances conspire that make it impossible for an owner to continue to keep their pet. If there's really no way around this heartbreaking problem, you could ask cat-loving friends or family if they can take care of your pet for you. Another alternative is to contact rescue organizations, but be aware that these shelters are often full to capacity, and can't take more animals in on demand. You may have to be placed on a waiting list until there is space for your cat.

One very definite and big no-no is to let your cat loose into the wild to fend for itself. It's incredible how many people still think it's either sensible or acceptable to do this. Pet cats that are kicked out are likely to become malnourished or even starve and be injured in fights or by traffic. Cats may be clever, but once they've been cared for by humans they thrive best in a home where they can receive practical care and affection.

❧ CAT OR KITTEN? ❧

You'd like a feline pet, but should you go for one that's fully grown, or opt for a baby? It's a tricky decision. These are the pros and cons:

KITTEN

For

- Ultra-cute.
- Young enough to adapt easily to new surroundings and other pets.
- You can watch your kitten develop from a little ball of fluff to a mature cat.
- House-training should be straightforward.

Against

- Needs lots of attention and can't be left alone for as long as an adult.
- Must have regular small meals until about six months.
- Kitty-play can be quite hard on furniture and carpets
- Even if house-trained, there may be the occasional accident.

CAT

For

- Is used to people and being handled.
- Won't wear you out with its hyperactive antics.
- Doesn't need as much full-on attention as a baby cat.
- Can be left on its own for several hours.
- Is usually already house-trained.

Against

- If your cat has come from another home, it may find it hard to settle.
- Probably won't take kindly to other pets in the house, especially other cats.
- Not as much fun to watch as a playful kitten.

CATS AND COMPUTERS 8
And why shouldn't I sit on top of the screen?
Get a flat one, if it bothers you.

WHY DOES MY CAT TRAMPLE MY LAP?

You've sat down with a coffee when you hear a "prrrpp," and up jumps your cat. She settles herself down on your lap, then gradually starts rhythmically kneading, trampling your lap with her paws, purring loudly all the while. She's in a blissful state, but it hurts when those sharp, needle-like claws pulse into your thigh. There's nothing for it, and you tip her off, only to discover that to add insult to injury, she's left a trail of dribble behind her.

Why do cats behave like this? It's because they are returning to that ecstatic state of kittenhood, when they were conditioned to knead their mum's belly to stimulate the flow of milk, and their dribbling is in anticipation of the delicious food to come.

So, when your cat treats you to a thorough lap-pummelling, don't take offense. She is demonstrating just how much she trusts and loves you. Just think of yourself as her mother-figure, and try to forget the pain.

WHY ARE CATS SO KEEN ON DARK, ENCLOSED PLACES?

Hands up who's ever searched the house for their cat, and found him happily curled in some cramped and, to human eyes, uninviting spot – in the space under a shed or garage, beneath the bottom drawer of a chest, or even crammed into a drawer that's already full.

But your pet is only making himself feel lovely and safe. The wide open spaces of the wild are dangerous. Any savage creature might see you – or kill you, even. Far safer to creep into a narrow, dark slit of a hiding place and stay there until the coast is clear. So, your sweater drawer is a fine substitute for a hole in a tree trunk, a burrow or a gap between a couple of boulders.

"Make fast your door with bars of iron quite;
No architect can build a door so tight
But cat and paramour will get through in spite."
Apollodorus of Carystus, Greek playwright,
c. 300–260 B.C.

LIVING WITH A DEAF CAT

Like humans, cats can be born deaf, or, more often, gradually lose some of their hearing as they grow older. You may not even realize that Maisie has trouble picking up sounds, until you make her jump one day because she didn't hear you coming.

A cat that becomes deaf suddenly may start to be unusually clingy, following her owner around constantly, or showing other unusual behavior patterns. You might notice her turning her head, as she struggles to pick up elusive noises.

It's all too easy to frighten a deaf cat, who's been robbed of her most vital alert system. A startled cat tends to lash out, and you might be bitten or scratched if you wake her suddenly, or be greeted with a venomous hiss until she realizes that you are friend, rather than foe.

You can help your deaf cat by:

• Alerting her to your presence through vibrations rather than sounds. Slamming a door, clapping your hands sharply, or patting the sofa or comforter where she's lying can create enough vibration to make her realize you're there.

- Use visual signals to communicate. You can call your deaf cat in from the garden with a flashlight, or get her attention at meal times by tossing a small object, such as a toy, into her line of vision.
- Some owners keep their pets in permanently when they realize they are deaf to protect them from the hazards of traffic. Another possibility is to confine Beethoven to a well-fenced garden. In case he does manage to roam further afield, fit him with a collar with a loud bell so that you can track him down, and consider marking his collar with the words "I'M DEAF," so that anyone he comes into contact with will understand why this crazy cat doesn't bolt when the car engine starts.

WHY WOULD A CAT BE READING A STREET MAP?

Perhaps because the cat in question – a silver tabby with square spectacle markings around its eyes – is actually Minerva McGonagall, deputy Headmistress of Hogwarts School in J. K. Rowling's *Harry Potter* series. McGonagall is tall, prim and wears square spectacles. She teaches Transfiguration, which is, she believes, among the most complex and dangerous magic taught at Hogwarts, and has mastered the art so well herself that she can turn into a cat at will.

She appears in the opening pages of the very first book, *Harry Potter and the Philosopher's Stone* when Mr. Dursley notices something peculiar – a cat reading a map. Or is it a trick of the light? In fact, it's Professor McGonagall, waiting for Albus Dumbledore, on the night that the baby Harry is brought to live with his aunt and uncle, Petunia and Vernon Dursley – the night that Harry's magical parents have been killed, by the wicked Voldemort.

ALL ABOUT ELDERLY CATS

Life expectancy has risen for cats, as for humans. It's no longer uncommon for a cat to reach the age of 18, and a few go on until they're about 20. Most cats slow down as they age, although the inner kitten is still there, and they'll have a quick little romp when they get the chance.

An elderly cat starts to look older. Methuselah's coat will show some gray hairs, and his fur will gradually become less glossy. His eyes will be less sparkling, his whiskers less perky and he won't prick his ears at the faintest sound. He'll probably spend almost the whole day dozing in a comfy spot, and when he does move he'll be slower.

Old cats become less graceful, and tend to topple off surfaces more easily. Methuselah may take the stairs one at a time, especially on the way up, if his old legs have become a bit creaky with rheumatism or arthritis. He won't spend so much time sniffing around in the garden, although he might still enjoy a daily stroll outside, especially if the weather is fine. And in summer you might find him sprawled in heavenly relaxation across a sun-warmed paving stone or bench. He probably won't rub around you as cravenly as he did when he was a younger chap, and he'll drop some of his more antisocial habits, such as furniture scratching, as well.

Scrawny senior cats are probably eating fewer calories because they're less active than before, but others tend toward a plump old age, and relish their meals as one of life's little pleasures.

Your elderly cat doesn't need any special care just because he's old. Make sure he has a warm and draft-free bed which he can get into and out of without a struggle. Switch him to a senior cat food if you wish, and let the vet give him the once-over every year so that any potential problems can be spotted early. Keep an eye on his teeth, as many mature felines shed teeth unnoticed. They also lose weight because they find it so hard to eat. Other than that, continue to give your pet your usual love and affection, and enjoy the remaining years that you have together to the full.

◦•❦ ARE CATS GOOD FOR YOUR HEALTH? ❦•◦

The Japanese certainly think so. Tokyo workers can ease the angst of the boss's unreasonable demands by visiting Cats Livin,' an animal-therapy center. For a mere 1,000 yen they can spend their lunch break stroking a houseful of friendly felines. Soothed and restored, the prospect of the afternoon office stint just doesn't seem so bad.

Stroking a cat is known to be super-calming, and can even lower the blood pressure, which may be why cat-owning heart attack patients have an improved survival rate.

Caring for a loving, dependent cat is a recipe not just for physical well-being, but for emotional and psychological health as well. Cats are particularly beneficial to older owners. Elderly cat fans make fewer visits to the doctor's office, and the time they spend simply watching their pet's antics can reduce anxiety and depression. Joining in with cat-play is even better, and owners who enjoy a good bout of chase-the-string find their cares and worries are forgotten, while they have fun with their cat.

Cats are responsive and affectionate companions. How can you feel lonely, when your feline friend waits for you to come home, and leaps into your lap the moment you sit down? What's more, cats need their people – and there's nothing better for people, than feeling needed. When the dinner bowl needs filling, and the fur needs brushing, there's an element of order and ritual to the day, which helps lonesome cat owners feel valued. Petting an appreciative, purring cat is not only a delight because that warm furriness feels so good. It also creates a warm glow of self-worth because it's so obvious that your pet relishes your attention.

Cats are even "prescribed" for good health. Whenever animal charities take docile trained cats, and dogs into hospitals, hospices and nursing home to give inmates a welcome dose of pet therapy, the patients' sense of wellbeing goes up, as their blood pressure goes down. Cats are a great source of good health.

What happened to the cat who
swallowed a ball of wool?
She had mittens.

·≋ *CATS* MAN'S CAT ·≋

Not surprisingly, Andrew Lloyd Webber, the composer of *Cats,* is a cat lover, and owns a rare Turkish Van "swimming cat" called Otto. The breed has large paws, which enable it to paddle around happily in the water, but it loves water so much that if it doesn't get the opportunity to take a dip in a lake, it'll look elsewhere for a cat-sized swimming pool. Breeder David Johnson owns 16 Vans, and warns that: "Given the chance they could well leap into the bath, and owners should be advised to keep their toilet lids down."

It's not just bathroom fittings that can lure a curious cat to go exploring, as Lloyd Webber discovered when working on the score of the follow-up to *The Phantom of the Opera.* While the composer's back was turned, Otto shimmied inside his digitized piano with disastrous results. Said Lloyd Webber: "Otto got into the grand piano, jumped onto the computer and destroyed the entire score for the new *Phantom* in one fell swoop." Experts couldn't retrieve the lost music, and Lloyd Webber had no choice but to start again – from scratch.

·≋ CALL A CAB! ·≋

What do you do when you're moving from one side of the continent to the other – with your pets? Bob and Betty Matas couldn't bear the thought of consigning their two cats to an icy journey in the hold of the plane, when they moved 2,500 miles from New York to Arizona. They hired a cab instead, and had the rear fit with carpet-lined cages, complete with litter boxes, so the cats could travel in style. The prospect of keeping the meter ticking the whole time was too terrifying, so they negotiated a flat fee with the delighted cab driver of a cool $3,000, plus expenses.

·≋ CAN I STOP MY CAT SCRATCHING THE FURNITURE? ·≋

You can but try. Scratching is one of a cat's basic instincts, and is one of a range of behaviors designed to mark territory. Perkins isn't sharpening his claws when he scratches. He's simply staking a claim to the surrounding area. And he definitely doesn't have an overpowering desire to destroy your new sofa, even if it feels that way.

It's in your furniture's interests to provide a couple of scratching posts for your pet, strategically placed around the home. Encourage him to use them initially by putting catnip or food treats at the base of the post. Make sure posts are at least three feet high, because Perkins needs to stretch out to his full length to do a thorough job of leaving his mark, and a shorter post just won't turn him on.

Another undesirable scratching habit is frantic clawing at the base of a closed door, often the bedroom, and usually when you are trying to keep the cat out. Try this – position a large scratching post between the door and its frame, pulling it into position as you close the door. With a bit of luck, Perkins will claw the post a few times instead of the carpet, then meander off to try his luck elsewhere.

☙ TOP PRICES FOR A MOUSER ☙

In out of the way, wild places, where cats were unknown, the mice population – and the price of a good mouser – could go through the roof. In the late 19th century, the newly affluent gold town of Deadwood, S.D., was desperate for cats. But none were to be had, until Phatty Thompson, a mule skinner and entrepreneur, rolled 'em in by the cart-load. He priced them, not by the puss, but by the pound, so the largest critters went for more than $30 – about $500 in modern currency.

☙ CAN CATS EAT CHOP SUEY? ☙

Cats crank up the "aaah" factor in TV ads when they're trained to perform cute tricks. And in the days before animal rights, performing cats were part of the circus scene.

- It may have taken six hours hard work a day, but the training paid off for the American owner who trained a cat to eat with chopsticks. After a single TV appearance, puss was swamped with lucrative job offers.
- Pure white, and with such good table manners, Arthur made his name delicately nibbling on cat food he'd scooped out of the can with his paw.
- Signor Capelli's Circus had a troupe of performing cats who could juggle, tightrope walk and perform on the trapeze. These 19th century marvels took their commands in three languages.
- George Techow spent three years training each of his cats to walk on their front paws and leap through flaming hoops.
- The Moscow State Circus had a team of cats that could fling themselves over obstacles, pull a carriage and play chess.

❧ CAN YOU KEEP IT SHORT? ❧

Haiku, the truncated Japanese verse form comprising a seven-syllable line sandwiched between two lines of five syllables each, lends itself to succinct observations about the unmistakable ways of cats. These two are in cat-speak.

The rule for today:
Touch my tail, I shred your hand.
New rule tomorrow.

***

Tiny can, dumped in
plastic bowl. Presentation:
one star. Service: none.

And this is the owner's lament:

Cat, fearless hunter,
Leaves "present" for me near door.
Next time I'll wear shoes.

❧ EXACTING STANDARDS ❧

Poppy hates being left alone. She cannot abide it when her people desert her. If it's just a couple of days, she'll almost cope. But no more. Definitely not.

We used to leave her for a week or more. The neighbors would come in twice a day to provide food and strokes, but it wasn't up to Poppy's rigorous standards of cat care, and she went AWOL, in search of some new owners who would look after her properly. We came back from holiday to find no cat, and the answering machine full –

"A lovely little cat keeps coming into our house. . . . bleep. . . . That little tortoiseshell is back again today. . . . bleep. . . . she's spending all day here now. . . . bleep. . . . she's wearing your phone number on her collar, but you never return our calls. . . . bleep. . . . If you don't want her, we'll take her!"

We raced around and retrieved our indignant cat from her new owners' sofa, but we couldn't risk it again. Now Poppy goes to a lovely cattery we found, where they have heated beds and hugs on demand. *Prrrrrrrrrr.*

❧ CRAZY CAT BELIEFS AND CONTRADICTIONS ❧

Take your pick from these ancient superstitions about cats. There's one to fit every situation.

- A cat jumping on to a sick person's bed in Germany meant the person was doomed to die. In other countries, it was the cat leaping off the patient's bed that meant the Grim Reaper was lurking on the threshold.
- In some times, and in some places, killing a cat has meant certain bad luck. In other times, and in other places, killing a cat has been the only sure way to keep the Devil at bay.
- If the cat leaves home, it takes the luck with it. Or, when a strange cat strolls in, it stops the bread from rising and ruins the fisherman's catch.
- Want to be invisible? Snack on the brains of a freshly killed cat. Make sure they're still warm, or the magic won't work.
- Weeds rampaging through your garden? A sure-fire cure for couch grass is to bury a live cat nearby.
- Taken a tumble? Reduce the bruising by eating cat poo. It works, honest.

◦❖ DRINKING AT THE CAT AND KITTENS ❖◦

The pub sign *The Cat and Kittens* refers to the large and small pewter pots that beer used to be served in.

◦❖ WHAT IS THE WORLD CAT POPULATION? ❖◦

There are more than 200 million cats on the planet, and a quarter of them are living in the United States, where there is one cat for every four people. Another 4.6 million live in Canada, and there are 12.5 million in Russia.

Britain has 7.5 million – a lot of cats for a small island – France has 8.5 million, Italy seven million, but Spain only three million.

◦❖ HOW TO CAST A CAT ❖◦

A radio competition asked listeners to send in pictures of pets that looked like stars. Most entrants were dogs, but two cats with star quality stood out.

* Izzy, a dead ringer for Warren Beatty.
* Sinead, who with her boggly eyes and protruding mass of hair, could easily be mistaken for cartoon character Marge Simpson.

Casting company Lookalikes-USA took the whole matter very seriously. "It is not unfeasible that we would get a call for an animal celebrity look-alike," said a spokesman. "The field is always expanding."

◦❖ WHAT HAPPENED IN SALEM IN 1692? ❖◦

A series of celebrated witchcraft trials took place in the Massachusetts port of Salem in 1692, and 19 people were hanged. The trials were founded on exaggerated witness reports, and confessions extracted under torture.

This example comes from the trial of Susanna Martin, who, it was testified, could bewitch cattle, had a wicked cast in her eye that could send people sprawling to the ground with one glance, and could arrange for a devil in the form of a cat to persecute anyone who crossed her. She was later executed, and the trial was reported by the Rev. Cotton Mather, one of the leading ministers in New England in the late 17th century.

Robert Downer testifyed, That this Prisoner being some years ago prosecuted at Court for a Witch, he then said unto her, He believed she was a Witch. Whereat she being dissatisfied, said, That some Shee-Devil would Shortly fetch him away! Which words were heard by others, as well as himself. The Night following, as he lay in his Bed, there came in at the Window the likeness of a Cat, which Flew upon him, took fast hold of his Throat, lay on him a considerable while, and almost killed him. At length he remembered what Susanna Martin had threatned the Day before; and with much striving he cryed out, "Avoid, thou Shee-Devil! In the Name of God the Father, the Son, and the Holy Ghost, Avoid!" Whereupon it left him, leap'd on the Floor, and Flew out at the Window. …

❀ WHAT MAKES A CAT FEEL STRESSED? ❀

Cats like consistency. Of course, they wish to come and go exactly as they please, but they like to have an underlying predictable routine. They are also very habituated to their own territory. Any event that disturbs their daily pattern or threatens their territory can make a cat nervous and stressed. These are common causes of feline angst:

- House moving is the classic cat-stressor. Being whisked away from all her well-known landmarks is very upsetting for a cat.
- Other cats arriving in – or leaving – the household, or even the neighborhood.
- Being bullied by another cat, or having an accident or illness can cause an upset.
- Any change in her owner's routine, related to a new job or new relationship, can make your cat feel ill at ease and wonder: What's going on?
- Changes in the family structure, from the arrival of a new baby to the departure of a noisy teenager, can be enough to disturb her equanimity. Guests can also make her feel anxious.

You can tell if your cat is feeling worried or stressed. She:

- Starts grooming herself much more often than usual, and there's an obsessive air to all that furious licking.
- Gives you the cold shoulder and won't respond to calls, strokes or other blandishments.
- Goes from being a happy wanderer to a housebound recluse, refusing to go out and rushing back in if you put her out of the door.
- Turns her nose up at her food – this can also indicate physical, rather than psychological, disturbance, and your vet should check her over.

◦❖ I MISS MY MOM ❖◦

Some cats – and you'll know for sure if yours is one of them – simply adore their owners. Like a fixated toddler, they'll follow their person around, anxiously watching in case they're about to be abandoned. This can happen when a cat who loves company has to be left alone for long periods – when her owner gets a new job, for example.

You can tell if Tinkerbell's overattached by the way she behaves.

- When you're away, she won't interact with other people and appears anxious and withdrawn.
- She won't relax or explore unless you're nearby.
- When you're on tap, she won't leave you alone and demands your constant attention.
- If you're away for any length of time she wakes frequently, grooms much more than usual, and may even chew or suck her coat or claws.

It's not an easy problem to tackle, as the only real solution is to make yourself more available to your stressed-out feline.

◦❖ THE KILKENNY CATS ❖◦

The saying "Fight like Kilkenny cats" comes from the following verse. The story goes that, during the 1798 Irish rebellion, Kilkenny was garrison to a troop of Hessian soldiers, who found cruel sport in binding two cats together by their tails and chucking them back and forth across a rope to fight, taking bets on the result. When an officer appeared, a soldier slashed off the cats' tails with his sword and the cats fled, leaving nothing but the bloodied tails, which the soldier explained by saying the fight had been so hard, that nothing but the tails was left.

There were once two cats of Kilkenny.
Each thought there was one cat too many,
So they fought and they fit,
And they scratched and they bit,
'Til, excepting their nails,
And the tips of their tails,
Instead of two cats, there weren't any.

❧ ENJOYING A TIPPLE ❧

Jerome K. Jerome, author of *Three Men In a Boat*, knew of a cat who supped from a leaky beer tap until she got drunk. Other cats on record had a fondness for porter, or brandy and water.

❧ ALEXANDER MCCALL-SMITH ON CATS ❧

Author of more than 50 books, including the bestselling *No. 1 Ladies' Detective Agency* mysteries featuring detective Precious Ramotswe, and *The Sunday Philosophy Club*, he was born in what is now Zimbabwe. McCall-Smith taught law at the University of Botswana before coming to the United Kingdom. He retired as professor of medical law at Edinburgh University in 2005.

McCall-Smith might seem to be more of a dog man than a cat lover, when he said: "I also hope for guidance on the profound question: with so many good, loyal dogs to love, why do we humans waste our affection on cats, who surely are less worthy of it?" It seems he's not entirely negative about cats, or at least about their finely tuned instincts.

In an interview with Jane Hutcheon in Botswana, he said:

"There's a nice cat over there. It's always a good sign when you see cats prowling around the street. That suggests that it's a good place to live."

Hutcheon: "Really?"

Alexander McCall-Smith: "Oh, yes, they can tell."

❧ INDEX ❦